A HISTORY OF THE SERVICES

OF

THE 17TH

(THE LEICESTERSHIRE) REGIMENT

MAJOR GENERAL JAMES WOLFE, 1759, (BY GAINSBOROUGH),
IN THE POSSESSION OF
MRS. HORACE PYM.

By the courtesy of the Editor of "The Connoisseur."

A HISTORY OF THE SERVICES

OF

The 17th

(The Leicestershire) Regiment

CONTAINING AN ACCOUNT OF THE FORMATION OF THE REGIMENT IN 1688, AND OF ITS SUBSEQUENT SERVICES, REVISED AND CONTINUED TO MARCH 31ST, 1912.

BY

LT.-COLONEL E. A. H. WEBB

(Formerly of the Regiment).

SECOND EDITION.

DEDICATED

TO

THE OFFICERS,

NON-COMMISSIONED OFFICERS

AND MEN

NOW SERVING, OR WHO HAVE SERVED

IN THE

LEICESTERSHIRE REGIMENT

(OLD 17TH).

THE 17TH

(THE LEICESTERSHIRE) REGIMENT OF FOOT

BEARS ON THE REGIMENTAL COLOUR

"*THE ROYAL TIGER*"

WITH THE WORD

"HINDOOSTAN"

AS A LASTING TESTIMONY OF THE EXEMPLARY CONDUCT OF THE REGIMENT
DURING ITS SERVICE IN INDIA FROM
1804 TO 1823;

"NAMUR, 1695;"

"LOUISBURG;"

ON ACCOUNT OF BEING ENGAGED IN THE MEMORABLE SIEGE;

"MARTINIQUE, 1762;" "HAVANNAH;"

ALSO THE NAMES

"AFFGHANISTAN," "GHUZNEE, 1839," AND "KHELAT,"

IN COMMEMORATION OF ITS DISTINGUISHED CONDUCT IN AFFGHANISTAN,
THE STORMING OF GHUZNEE, AND THE CAPTURE OF KHELAT,
IN THE YEAR 1839,

ALSO

"SEVASTOPOL," "ALI MASJID,"

"AFGHANISTAN, 1878-79," "SOUTH AFRICA, 1899-1902,"

"DEFENCE OF LADYSMITH."

PREFACE.

THIS regimental history is a revision and continuation up-to-date of the printed Historical Record of the 17th Foot, published in 1848, by the late Richard Cannon of the War Office.

Whilst the original construction has throughout been partly preserved, a considerable addition has been made, detailing moves of the regiment, its numerous engagements, and other subjects of interest, the particular aim having been to compile mainly from original despatches and from the "London Gazettes," and to avoid everything of a political nature, or such as might bear on any political controversy.

It must always be a matter of the keenest regret that the official correspondence books of the regiment and other valuable documents have been on two occasions lost, viz. :—

Firstly, in the Helder Expedition (1799), when the books were lost of both battalions.

Secondly, the 2nd Battalion having been disbanded in 1802, the official books of the 1st Battalion were again lost in the wreck of the transport "Hannah" off the coast of Bombay, in March, 1840, when all baggage, including the mess plate, had to be thrown overboard to lighten the ship. The grievous loss of the regimental papers has debarred an insight being obtained into such old regimental traditions as are known to have existed.

The loss, for instance, of any documentary evidence as to the origin of the "Green Tiger" on the regimental colour, the motto "Veni et Vici," and the black line in the officers' silver lace, is irreparable. The regimental "Digest of Service," dated Hull Citadel, 1st May, 1824 (belonging to the depot, and under present regulations at Lichfield), gives no particulars of such details, and there is no doubt that, had the regimental papers been preserved, a careful examination of them would not only have considerably lightened one's labours but would have given interesting particulars of the internal life of the regiment, which are not obtainable from the examination of public documents, however carefully they may be searched.

By the courtesy of His Excellency The Spanish Ambassador, enquiries have been made through the Spanish Embassy for any information that it might have been possible to obtain of the regiment during its twenty-four years' sojourn at Minorca (1725-1749), but unfortunately without success, from which it would appear that, beyond a detachment of it having been sent to Gibraltar in 1727 (to assist in the defence of that fortress against the Spaniards), its stay at Minorca was otherwise uneventful.

In accordance with custom up to the year 1751, the regiment has been designated in these pages by the name of its colonel, regimental numbers not having been allotted to line regiments until July 1st of that year, though on reference to the old official books, it will be seen (from a continued use of the Colonels' names for many years later) that it was apparently as difficult at that period to get used to the regimental numbers as it has since been to discontinue them on their abolition, 1st July, 1881.

My first debt of gratitude is due to the late Mr. Milne of Calverley House, Calverley, near Leeds, who, for some years before his death, had attained the high position of being considered the greatest authority of the day on all matters referring to army dress and equipment, and also to regimental colours. His useful help has often been eagerly sought by many regiments throughout the service, as can be seen by the numerous publications of Regimental Records to which he has contributed much valuable information.

I am much indebted to the authorities at the Public Record Offices in London and Dublin, to the India Office, and Prince Consort's Library, Aldershot, and to Mr. Cary (Librarian at the War Office) for his genial aid always so readily given.

Also (amongst old brother officers) to Major-Generals Tompson, C.B., and Utterson, C.B. (Colonel of the regiment), Colonel Mansergh, Lieut.-Colonels J. G. Anderson, F. S. S. Brind, F. C. Maisey (Indian Army, retired), and, of those now serving, to Lieut.-Colonels L. C. Sherer, H. L. Croker, Majors Blackader and Stoney Smith, and Lieut. Panton for their ready and generous assistance. In mentioning Lieut-Col. Maisey and Major Stoney Smith (Hon. Secretary to the Regimental History Committee), it is difficult to express sufficient appreciation of the valuable help they have given.

The following ladies and gentlemen have very kindly assisted with information, or in lending documents and portraits, and in interesting others :—

The Misses M'Kinstry, Bangor, North Wales (whose father and eldest brother each commanded the 2nd Battalion of the regiment).

Mrs. Urquhart Mosse, Jersey, Channel Islands.

Captain F. A. Wetherall, R.N. (retired), Folkestone.

Lieut.-Colonel A. R. Savile, late 18th "Royal Irish" Regiment, Hastings.

L. Peevor, Esq., South Hampstead, London, N.W.

W. K. Fayle, Esq., J.P., Birr, Ireland.

J. C. Leask, Esq., Newcastle-on-Tyne.

F. W. Thomas, Esq., India Office.

Mrs. Millbank Leslie-Melville, Collessie, Fife.

The Misses Mauleverer, Hastings (whose father served in the regiment in the Afghan Campaign, 1839).

C. Dalton, Esq., F.R.G.S., London, S.W.

Major G. H. Lawrence, 1st Battn. East Lancashire Regt.

I must also gratefully acknowledge the unsparing pains taken by Mr. H. Tregoning, of Ashford, Middlesex, in obtaining with such good results, as an amateur, the photographs he has taken of regimental ornaments, &c., and for his useful help.

I am also indebted to the following firms for useful information relative to the uniform and appointments of the regiment at early periods, viz.: Messrs. Hawkes and Co., 14, Piccadilly, W.; Messrs. Jennens and Co., 56, Conduit Street, W.; Messrs. Herbert and Co., Bedford Street, Covent Garden; the Goldsmiths' and Silversmiths' Company, Regent Street; and Mr. A. D. Baldwin, Duncannon Street, London.

E. A. H. Webb,

Lieut.-Colonel, Retired.

(formerly of the 17th Regiment).

March 31st, 1912.

The following abbreviations occur as references, viz.: W.O., signifying "War Office"; C.O., "Colonial Office"; P.R.O., "Public Record Office"; R.U.S.I., "Royal United Service Institution."

A HISTORY OF THE SERVICES OF THE 17TH (THE LEICESTERSHIRE) REGIMENT.

CONTENTS.

CHAPTER I. 1688—1697

YEAR. PAGE

CHAPTER III. 1708—1757

CHAPTER X. **1847—1856.**

CHAPTER XIII. **1882—1898**

CHAPTER XIV. **1899—1910**

APPENDIX.

Succession List of Colonels.

Biographies of the Colonels.

"The Truth about the Siege of Londonderry in 1689." (A vindication of Colonel Solomon Richards.)

An Account of the Battle of Sheriffmuir, 1715. By Major-General Wightman, Colonel of the regiment.

Regimental Plate, Pictures, &c.

Claim for compensation for losses sustained by the 17th Regiment in the wreck of the transport "Hannah" in 1840.

The Regimental Chapel, St. Martin's Church, Leicester.

Three Presentation Cases.

Regimental Music.

LIST OF PLATES.

ILLUSTRATIONS IN THE TEXT.

Errata.

Page 13, line 16, for "Lieutenant" read "Capt.-Lieut. Evans."

Page 31, line 19, for "13th" read "31st."

Page 226, line 13, for "Thorald" read "Thorold."

"*A Regiment, great in history, bears, so far, a resemblance*
"*to the Immortal Gods, as to be old in power and glory—yet to*
"*have always the freshness of youth.*"—*Kinglake.*

A HISTORY OF THE SERVICES

OF

THE 17TH

(THE LEICESTERSHIRE) REGIMENT.

CHAPTER I.

HOME SERVICE AND FLANDERS.

1688–1697.

1688

IN the autumn of 1688, when the adoption of pernicious counsels by the Court had given rise to the preparation of an armament in Holland to support the British people in the preservation of their religion and laws, and King James II. began to entertain apprehension for the permanence of his government, His Majesty issued commissions for adding to his regular army five regiments of cavalry and seven of infantry, including two corps formed of men who had quitted the Dutch service; and of these twelve regiments, the 16th and 17th Regiments of Foot in the British line are the only remaining corps.*

The regiment which eventually became the 17th (in 1751) was raised in London and its immediate vicinity, and the colonelcy was conferred on Solomon Richards, by commission, dated 27th September, 1688.

* The regiments raised in 1688, by King James II., were commanded by the following officers:—HORSE.—The Earl of Salisbury, Marquis de Miremont, Viscount Brandon, Henry Slingsby, and George Holman. FOOT.—John Hales, Roger McEligot, Archibald Douglas, Solomon Richards, the Duke of Newcastle, Colonel Gage, and Colonel Skelton.

The following is a copy of the warrant for raising Captain Solomon Richards' company of foot :—

" To Our Trusty and Welbeloved Solomon Richards "
" Esqre, Captain of a Company, in Our Regt. of ffoot,"
" whereof He himself is Colonel."

" These are to authorise you, by Beat of Drumm or "
" otherwise, to Raise Voluntiers for your own Company in "
" Our Regt. of ffoot, whereof you are Colonel, which is to "
" consist of Sixty Private Soldiers, Three Serjeants, Three "
" Corporals, and Two Drummers, to be Mustered as each "
" Soldier or Non-Commission-Officer shall be respectively "
" produced unto the Commissary General of the Musters, his "
" Deputy or Deputys, and to be allowed for one Month before "
" the Day of such Muster, and, for such further time as They "
" shall be Mustered by you. And, you are to Appoint such "
" Person or Persons as you shall think fit to receive Arms "
" for the said Voluntiers out of the Stores of Our Ordnance. "

" And We do hereby Require all Magistrates, Justices "
" of the Peace, Constables, and other Our Officers, whom it "
" may concern to be Assisting to you, in Providing Quarters "
" and otherwise, as there shall be Occasion. Given "
" at Our Court at Whitehall, the 27th day of September, "
" 1688, &c."

" By His Mats Command,"

" WILLIAM BLATHWAYTE."

The like Orders for Raising the Companys follg. in the said Regiment.*

The following is the first list of officers of

COLONEL RICHARDS' REGIMENT OF FOOT.

(All commissions bear date 29th September, 1688, except those marked otherwise.)

Colonel Solomon Richards, 27th September.
†Lieut.-Colonel Sir Matthew Bridges, 27th September.
Major John Jones.

* W.O. 26, Book 6.
† Sir M. Bridges was knighted by King James II., in honour of the birth of the Prince of Wales, he (Bridges) happening to be in Windsor at the time.

Captains: Joseph Taylor, Dan Wicherley, Sam Bridges, Hundgate Lewis, John South, George Capstack, Sir Herbert Whaley, Edward Probee, Richard Smith, Richard Tucker, Dugall Campbell (Grenadier Company).

Lieutenants: Dennis Pearson (Capt.-Lieutenant), George Withers, Henry Paget, Henry Byland, John Campbell, Ben Alsop, Richard Driver, Robert Aske, Wm. Maydman, Anthony Beverley (30th September); Wm. Mouse and John Douglas (Grenadier Company).

Ensigns: John Bradshaw, Nathaniel Taylor, Thos. Morris, Jno. Pierre Desbordes, Adam Lesney, Ebenezer Bucknall, Theodorus Smith, 30th September; Hugh Moffat, Christopher Russell.

Adjutant John Evans, 4th October.

Quarter-Master Thos. Browne, 4th October.

Great success attended the efforts made to procure men for completing the ranks of the regiment, and in three weeks after the letter of service for its formation was issued, it was embodied, armed, and clothed. It was composed to a great extent of men who had entered the army at the augmentation in 1685, and had been discharged after the suppression of the Duke of Monmouth's rebellion. The regiment was speedily reported fit for duty, and on the 23rd of October, orders were received for four companies to march to Colnbrook and Longford, four to Staines and Egham, and five to Windsor, Datchet, and Slough; at the same time two companies were directed to mount guard at the Castle at Windsor: thus were the first duties of the regiment those of a guard to the Royal person.

On the 29th of October, the quarters were changed to Maidenhead, Datchet, and Windsor; and on the 7th of November, when the Prince of Orange had landed in Devonshire, the regiment received orders to march to Greenwich and Deptford, to be in readiness to protect the establishments in the vicinity of those places, and to aid, if required, in the preservation of the peace of the Metropolis, leaving one company for duty at Windsor Castle.

On the 2nd of December, it was directed to march to Holborn, and, in the same month, to Sudbury (Suffolk), with detachments at Molford and Lavenham.

The clothing and equipment for the army, except the Ordnance Corps, were from the earliest period, up to 1855, supplied by the Colonel, who received a fixed sum for every man on the establishment, which was termed "off reckonings,"* and any balance remaining, after paying the cost of the articles, was a source of emolument. The first regulations on this subject were issued this year by King James II.†

On the 31st December, the regiment was reformed, with the addition of a chaplain, Solomon Foley, and chirurgeon (surgeon), James Blean.

1689

An Order, dated 3rd of January, directed Colonel Richards' Regiment to march, on the 11th, in two divisions, from Sudbury to Liverpool, Ormskirk and Walton.

The sum voted for the expenses of the regiment, this year, including clothing, was £16,145 3s. 4d. *All field officers, at this period, had, in addition to their field duties, to look after a company.*

The payment of levy money for each recruit raised was 40 shillings, and the daily pay was as follows:—

	£	s.	d.
Colonel: as Colonel, 12s.; as Captain, 8s.	1	0	0
Lieut.-Colonel: as Lieut.-Colonel, 7s.; as Captain, 8s.		15	0
Major: as Major, 5s.; as Captain, 8s.		13	0
Captain		8	0
Lieutenant		4	0
Ensign		3	0
Adjutant		4	0
Quarter-Master		4	0
Chirurgeon (Surgeon)		4	0
Surgeon's Mate		2	6
Chaplain		6	8

Pay of one company.

3 Sergeants at 1s. 6d. each; 3 Corporals, and 2 Drummers at 1s. each; 60 Privates, at 8d. each.

* "Off Reckonings," a specific account, so called, which existed between the Government and the Colonels for the clothing of their men.

† Old Pimlico Regulations.

The regiment to consist of 13 companies (including one Grenadier Company), of 60 men in each. Total, 780, besides officers.

The following was the weekly subsistence allowed to a foot regiment :—

	s.	d.
To a Lieutenant	14	0
,, an Ensign	10	6
,, a Sergeant	6	0
,, a Corporal or Drummer	4	6
,, a Private	3	6

The events which now followed in rapid succession occasioned the flight of King James to France, and the services of the regiment were transferred to the Prince and Princess of Orange, who were elevated to the throne by the title of King William and Queen Mary, in February, 1689.

In Ireland, the army adhered to the interest of King James ; but the Protestants of Inniskilling and Londonderry embraced the principles of the Revolution, and wrote to King William for assistance to enable them to preserve those places in his interest. Colonel Cunningham's (9th) and Colonel Richards' (17th) Regiments were directed to proceed to Ireland, to support the people of Londonderry ; and the two corps sailed from Liverpool on the 3rd of April. Contrary winds forced the transports to anchor at Hoylake ; but they again put to sea on the 10th of that month, and on the 15th arrived in the vicinity of Londonderry.

By date of commission (27th September, 1688), Colonel Richards was senior to Colonel Cunningham (31st December, 1688), but for some unknown reason the supreme command of the troops was given to Colonel Cunningham, as the officer commanding *the senior regiment.*

Colonel Cunningham, having King William's orders to follow the directions of Colonel Lundy, the governor of Londonderry, immediately acquainted him of the arrival of the two regiments, and solicited orders. The governor directed him to leave the regiments on board, and to come with several other officers to the town, where a council of war was assembled. The governor, having secretly resolved to give up the town to King James (who had arrived in Ireland, with a body of troops from France), gave the council a false statement of its condition, representing that there were not provisions for more than a week or ten days, and his assertions induced the council to decide that it would be imprudent to land the two regiments. The inhabitants were so enraged at the conduct of their governor that they determined to depose him, and sent, offering the government to Colonel Cunningham, who replied that, " being himself commanded by the King to obey the governor, " he could not receive any application from persons who " opposed that authority." The governor escaped from Londonderry in disguise, and the inhabitants made a most gallant defence under the Rev. George Walker and other leading men of the town.* (See Appendix, pp. 288–303).

Colonel Richards, at the council of war, had strongly protested against the two regiments leaving Londonderry, " because he looked on deserting that garrison, not only as " quitting the city, but the whole kingdom."† He was, however, overruled, and the two regiments returned, sailing from Ireland on the 19th of April. King William was so displeased at the state of the fortress not having been sufficiently investigated, and that the governor's suggestions had been yielded to when there was reason to doubt his integrity, that he deprived Colonels Cunningham and Richards of their commissions. The Committee of the House of Commons, which assembled to investigate the circumstances of the two regiments returning, absolved Colonel Richards from blame, finding that he, having to abide by the decision of Colonel Cunningham,‡ had, practically, no voice in the matter. Colonel Richards, however, was not employed again.

The regiment, on return from Ireland, marched to Chester, arriving 24th April.

The colonelcy was conferred on Sir George St. George, by commission dated 1st of May, 1689.

An order, dated 27th May, directed Sir George St. George's Regiment to march to Nantwich, Wrexham, Malpas, and Whitchurch. On the 2nd July it received orders to march to

* Cannon's Record, 9th Regt., Also John Mackenzie.
† Walker's " Siege of Derry."
‡ Siege of Derry, John Hempton, pp. 381–2.

Carlisle, and a letter from Sir John Lowther to the Earl of Shrewsbury, dated 11th August, shows that it had then left Carlisle for Keswick.*

1690

A marching order, dated 3rd April, directed the recruits of the regiment to march to Chester, to embark for Ireland, the regiment furnishing a draft of 280 men for transfer to the Earl of Drogheda's Regiment quartered there. The headquarters followed later to Chester, and a Route, dated 2nd December, directed the headquarters and five companies to march to Barnet, with detachments at Chester, Hadley, and Whetstone.

1691

On the 15th January, the regiment marched to Hull, detaching in October two companies to Reading, which moved later to Portsmouth, and in November three companies were detached to Berwick and one to Tynemouth.

1692

By the 18th February, the detachments at Berwick and Tynemouth had rejoined headquarters at Hull.

1693

The regiment proceeded early in November to a large standing camp for about 20,000 troops, which had been assembled between Petersfield and Portsmouth. Luttrell relates that His Majesty had resolved to send twelve foot regiments of this force to Flanders, and on the 11th November, Lord Cutts proceeded to Hampshire to inspect them before embarking.

On the 28th December, the regiment was reviewed by King William in Hyde Park, in brigade with Lloyd's (5th), Hamilton's (18th), and Ingoldsby's (23rd) Regiments, all of which were marching to embark at the Red House (on the Thames) for Flanders.†

* Calendar of State Papers—Domestic Series, "William and Mary."
† Luttrell's Diary.

The strength of regiments on the English establishment, from 1693 to 1696, is shown as 44 commissioned officers, 104 non-commissioned officers, and 780 men : a total of 928.

The following are peculiar instances of awards by courts-martial at this period, the crime of desertion having been invariably met by the penalty of death.

As the result of a court-martial held on four soldiers (of another regiment) for desertion, the following was the approved sentence, dated Whitehall, 3rd June :—

" The Queen's pleasure (in the King's absence, in " Holland) is that she will have only one of the four to dye. " M., having been sentenced to death, to have the Liberty " of Casting Lotts, with the three others, notwithstanding " the directions that have already been given for M.'s " execution."

Two sentences by a court-martial held at Portsmouth on the 21st July, 1693, for the trial of two other soldiers, of the same regiment, decreed that one, for desertion, " be shott to death," whilst the other, for intending to desert (being in company with the above, but returning to his regiment), was sentenced " by plurability of voyers, " to Runn the Gauntlett five times through 600 men, with " two days' Intermission 'twixt each time of running."*

1694

Flanders, 1694.

By the 13th January, the transports that embarked the four regiments for Flanders had arrived at Ostend,† and here Sir George St. George's (17th) was quartered until the spring.

After the severe loss sustained by the confederate army in the Netherlands in 1693, King William returned to England, at the end of the campaign, and arrived back in Holland about the 10th May, this year. On the 17th May, Sir Henry Bellasis received orders to march next day, with the garrisons of Bruges, Ostend, and those quartered on the canal of Neuport, to form a camp by Ghent, numbering in all 19 battalions. Arriving at Ghent on the 19th, they encamped at Marykirk, where they were joined by the garrison of Ghent, and also the

* W.O. 26, Book 8, pp. 77, 96, 97.
† Luttrell's Diary.

regiments of St. George (17th) and Fredk. Hamilton (18th).* On the 28th, the force marched to the general rendezvous by Louvain, and encamped with the right at the King's Quarter at Bethlehem, and the left towards Louvain.

On the 31st May, the King reviewed all the English infantry that had come to the camp, in the presence of the Electors of Bavaria and Cologne, and on the 3rd June, took up his residence at the cloister of St. Hertogendale, where St. George's (17th), with Granville's (10th), Tidcombe's (14th), Leslie's (15th), and Castleton's and Lowder's Regiments were formed in Brigade, under Brigadier-General Stewart.† The regiment took part in the operations of the army commanded by the British monarch—performing many long and toilsome marches in Flanders and Brabant; but it had no opportunity of distinguishing itself in action, and in the autumn it returned to the port of Ostend, where it passed the winter in brigade with Mackay's and Graham's Regiments.*

Precedence of the first fourteen old British infantry regiments, serving in the Low Countries in 1694, from which it will be seen that the 17th Foot then stood thirteenth in order of precedence.

1. The Royal.
2. Colonel Selwyn's.
3. Major-General Churchill's.
4. Colonel Trelawney's.
5. Colonel E. Lloyd's.
6. Royal Regiment of Fuziliers.
7. Sir Bevil Granville's.
8. Colonel Richard Brewer's.
9. Colonel Tidcomb's.
10. Sir James Lesley's.
11. Colonel James Stanley's.
12. Colonel Francis Collingwood's.
13. Sir George St. George's.
14. Colonel Frederick Hamilton's.

The Royal Warrant, settling the precedence of the above-named regiments was signed at Roosbeck on the 10th June, 1694.‡

* D'Auvergne. † Cannon's Record, 15th Regt.

‡ Dalton, Vol. IV., p. 291. It should be noted, that all names are spelt according to the authorities quoted.

1695

In the spring of this year, Colonel Sir George St. George obtained His Majesty's permission to exchange with Colonel John Courthorpe, to a newly-raised regiment, which was afterwards disbanded, Colonel Courthorpe being appointed from the 1st of May.

On the 26th May, our garrisons in Flanders marched to take the field, and Colonel Courthorpe's (17th), with several other regiments, had orders to march to Dixmude to form a camp under Major-General Ellenberg.

On the 7th June, the Duke of Wirtemberg arrived for the purpose of making a diversion in favour of the main army, and after reviewing the troops and dividing them into four brigades, he encamped before the Kenoque, a fortress at the junction of the Loo and Dixmude canals, where the French had a garrison. The regiment, commanded by Colonel Courthorpe, was in the 2nd Brigade, with Brewer's (12th), Tidcombe's (14th), and Leslie's (15th), under Colonel Sir James Leslie, and took part in the capture of several outposts belonging to the fort, and its grenadier company was engaged on the 9th of June in driving the French from the entrenchments and houses near the Loo Canal. This brigade had twenty men killed and wounded before the Kenoque.* The total loss of the British engaged was 587 men killed and wounded besides officers.†

While the regiment was before the Kenoque, King William invested the strong fortress of Namur, and Courthorpe's Regiment and several other corps marched to join the covering army, under Charles Henry of Lorraine, Prince of Vaudemont. Against this army Marshal Villeroy advanced with a French force of about seventy thousand men ; and the Prince, not having above thirty-six thousand men under his orders, withdrew to the vicinity of Ghent.

The regiment was subsequently employed in operations to protect the maritime and other towns of Flanders, and to cover the troops carrying on the siege of Namur ; and after the surrender of the town it was selected to relieve one of the corps which had suffered severely in the siege, and to take

* Captain Knight's History of "The Buffs." † D'Auvergne.

part in the operations against the castle. The regiment arrived at Namur on the 11th of August, and took its turn of duty in the trenches and in all services connected with this great undertaking; it had several men killed and wounded, and on the 16th of August Captain Hart was killed in the trenches.

When Marshal Villeroy approached at the head of a numerous army to raise the siege, Courthorpe's Regiment was in position at the post of St. Denis, where it was expected that the most vigorous exertions of the enemy would be made. The French not hazarding an engagement, the regiment was one of those selected to take part in storming the outworks of the castle on the 30th of August.

Some writers give this date as the 20th of August, but the "London Gazette," in publishing a despatch from Lord Cutts, dated Namur, 1st September, 1695, shows the date of the assault as Tuesday, August 30th.

Siege of Namur.

The following were the dispositions for the assault on the castle and citadel of Namur:—

Lord Cutts, with 3,000 English, was to attack the counterscarp and breach of the Terra Nova; the Count de Rivera (Major-General in the Spanish service) was to attack the breach of the Cohorn and that part of the line of communications next the Terra Nova, with 3,000 Spaniards and Bavarians, whilst Major-General La Cave, with 2,000 Brandenburghers, was to attack, on the right of Count de Rivera, the upper point of the Cohorn; 2,000 Dutch, under Major-General Swerin, being directed to attack the Casotte, and at the same time a colonel (Count Marsilly)* was to attack the lower town with 600 men. The signal was to be a considerable quantity of powder blown up, at the Old Battery, near the Brussels Port. Four sergeants and sixty men were to take part in the forlorn hope. These were to be followed by the grenadiers of the Guards, the remainder of the grenadiers of other regiments following, making in all 700, whilst 300 grenadiers were ordered to attack the line of communications. Courthorpe's (17th) and Mackay's Regiments were ordered in support, and Hamilton's (18th) and Buchan's were in reserve. The last three regiments were, however, directed to take up a position at the Abbey of Salsine, with orders to march

* Col. Davis's History of the 2nd "Queen's."

immediately after the signal was given, and to form up in rear of Courthorpe's Regiment, to receive further orders. About mid-day the signal for the assault was given, when the grenadiers rushed forward, under a heavy fire from the castle, to storm the breach of the Terra Nova, and were supported solely by Colonel Courthorpe's (17th), with drums beating and colours flying, but owing to a mistake in the signal the three regiments in reserve did not move forward in time, and the assailants were overpowered by superior numbers.* Courthorpe's Regiment advanced in gallant style, but was assailed by a storm of bullets which nearly annihilated it. Colonel Courthorpe was killed, Lieut.-Colonel Sir Matthew Bridges severely wounded, and 11 officers and 250 soldiers of the regiment were put *hors de combat* in a few minutes.

A despatch, in the "London Gazette," dated Namur, 1st September, 1695, states that the British troops "advanced "with great order and resolution, and had carried the top of "the breach, through a great deal of fire on all sides, but "found the enemy so advantageously posted and entrenched, "with the ground on their own side so very bad, that they "could not advance in any form, and were forced to retire." Upon the arrival, however, of the three regiments of the reserve, the attack was continued, and eventually, after a desperate resistance, proved successful.† The fighting lasted until the evening and, although the original design of carrying the castle and outworks failed, the allies had secured a position nearly a mile in length. The British loss this day was computed at 1,400 killed and wounded, and the total loss of the allies at under 9,000 men; the loss of the besieged, during the siege, being estimated at about 6,500.

The regiment had Colonel Courthorpe, Captain Coote, Lieutenant Evans, and 101 sergeants and rank and file killed; Lieut.-Colonel Sir Matthew Bridges, Captains Edward Wolfe‡ and Du Bourgnay, Lieutenants Desbordes and Ashe, Ensigns Fonsubran, Eyres, and Dennis, and 149 soldiers wounded.

King William was pleased to confer the colonelcy of the regiment from the 1st September on the Lieut.-Colonel, Sir Matthew Bridges, who had evinced great gallantry on this occasion.

* D'Auvergne. † Hamilton's Guards, Vol. I.

‡ Captain Edward Wolfe was grandfather to Major-General James Wolfe, the hero of Quebec. Beckles Willson, p. 2.

D'Auvergne says of Colonel Courthorpe: "No gentleman "ever fell more generally lamented than Colonel Courthorpe "did on this occasion, giving all possible hopes of an extra-"ordinary man in the military art if he had lived."

The MSS. Records of British Regiments, at the Royal United Service Institution, give the following:—

An instance of bravery occurred during this campaign. Captain Withers (17th), being posted in a chateau with only six men, stood against Villeroy's whole army for some hours, and when he saw they were preparing to storm him, he felt himself obliged to beat the chamade,* on which he had the same terms granted, and was better treated than those who had surrendered without firing a shot.

In the list of widows of several officers killed and died of wounds between 1689–95, a gratuity of £20 was awarded to Mrs. Mary Evans, widow of Lieutenant Evans, of Courthorpe's (17th) Regiment, killed before Namur. At the close of the war (1697) all such gratuities became pensions, which mostly terminated at His Majesty's death in 1702.†

Preparations were made for a second assault of the works of Namur, which was prevented by the surrender of the garrison. Sir M. Bridges' (17th) Regiment remained a short time near the captured fortress, and afterwards marched to the opulent city of Bruges, where it passed the winter.

The following advertisement from the "London Gazette," dated, March 16th, 1695, shows that sums approximating ten guineas a head were offered for the apprehension of deserters at this period.

> "Deserted from Captain Taylor, in the Honourable Sir Mat. Bridges' (17th) Regiment, Roger Bromeley, a tall black man, with black lank hair, slow speech, formerly a trooper;"
>
> "Alex. Scofield, a tall black man, wears a black wig, pock-broken, and snuffles very much;"
>
> "Thurston Walker, a corporal, a little well set man, with light hair, all of Lancashire. Whoever secures them, and gives notice to Captain Taylor at the Cabinet, in St. Paul's Churchyard, shall have 30 guineas reward, or proportionable for each. Or if they be at Sea, good Seamen in their places, and reasonable charges."

* This word often occurs about this period and later, signifying the beat of a drum, or sound of a trumpet, inviting to a parley.

† Dalton, Vol. VI.

1696

Early in the spring, the regiment was joined by a numerous body of recruits from England, and on the 12th of May it marched from Bruges to Marykirk, and was afterwards encamped along the canal towards Ghent. It was formed in brigade with the 1st Battalion " Royals," Churchill's " Buffs," Fairfax's (5th), and Hamilton's (18th) Regiments, under Brigadier-General Selwyn, and served the campaign with the army of Flanders under the Prince of Vaudemont; but no general engagement occurred, and in the autumn the regiment marched into quarters at Bruges.

1697

On the 13th of March, the regiment quitted its quarters at Bruges, and was afterwards stationed a few weeks in villages between Brussels, Vilvorde, and Malines; it was subsequently formed in brigade with a battalion of the Royals, Fairfax's (5th), and two regiments in the Dutch service, under Brigadier-General the Earl of Orkney; and it took part in the operations of the army of Brabant, under King William, until hostilities were terminated by the Treaty of Ryswick, and the British monarch saw his efforts to preserve the liberties of his country and balance of power in Europe attended with complete success.

The regiment was now detailed for home service, and sailed in three transports, four companies on arrival landing at Cork, and seven at Kinsale on the 20th December, whilst the ship conveying two companies had broken down in Plymouth Sound.

SHERIFFMUIR MEDAL, 1715.

CROSS OF THE ORDER OF THE "DOORANEE EMPIRE."

LOUISBURG MEDAL, 1758.

CHAPTER II.

IRELAND, HOLLAND, SPANISH NETHERLANDS, ENGLAND, PORTUGAL AND SPAIN.

1698–1707.

1698

THE breakdown of the ship, with two companies of the regiment on board, elicited the following marching order, dated January 8th, 1698 : " It is His Majesty's pleasure that " the two companies of Sir M. Bridges's Regiment, now in " Plymouth Sound, do land forthwith, and march to Liskeard, " Saltash, and Lostwithiel, proportionately, and remain there " until the ship shall be in a condition to receive them on " board again, in order to prosecute the voyage to Ireland."

On the 16th March, 12 companies of the regiment were ordered to march to the town and fort of Kinsale, " to keep " guard there " until further orders, and one company was detached to Sherkin Island and Skibbereen until the 9th July, when it rejoined headquarters.

An order, dated 28th July, directed the regiment to march from Kinsale to Dublin, leaving Cork 6th August, " and " resting Sundays."

On August 25th the regiment was posted as follows :—

Four companies to Belfast, three to Carrickfergus, one to Lisburn, one to Larn and Isle Magee, one to Fall and Malone, one to Hollywood, Cumber and Castlerea, one to Newtown and Bangor, and one to Donaghadee, Ballywalter and Ballyherbert.

On October 10th it supplied further detachments to Dobsland and Kilrout. (Irish Marching Order Book.)

1699

In compliance with the proclamation that had been issued, revising the establishments of regiments in Ireland at this period, Sir M. Bridges' (17th) Regiment was now to consist of 11 companies, mustering 37 officers, 22 sergeants, 11 drummers, and 396 privates—488 of all ranks.*

A marching order, dated 31st July, directed a re-distribution of the regiment in several small detachments, and on the 12th September the company at Strabane alone was distributed between Castlederg, Ballymagunge, Claudy, and Drumgally.

1700

On the 25th July, an order was issued to Colonel Sir M. Bridges, Knight, to march his regiment from their present quarters to Dublin.

1701

A marching order, dated 2nd June, directed the regiment to march from Dublin to Cork.

The decease of Charles II., King of Spain, on the 1st of November, 1700, was followed by the elevation of the Duke of Anjou, grandson of Louis XIV., to the throne of that kingdom, in violation of existing treaties; and war being resolved upon, King William issued a Royal Warrant, dated 29th May, 1701, directing 12 battalions, of 12 companies each (49 soldiers to each company), with their equipage, to be despatched to Holland as soon as possible.† Sir M. Bridges' (17th) being included in this force, the regiment embarked at Cork on the 15th June, and sailed for Holland, where it was placed in garrison at Gorcum. On the 3rd September, included in the above force, it was reviewed by King William III. on Breda Heath, where the regiments were encamped for the occasion about two miles from the town, and on the following day returned to their respective stations.‡

:olland, 1701.

* Dalton, Vol. IV., p. 216.
† W.O. 26, Book 11.
‡ "London Gazette."

1702

The campaign this year commenced on the 9th March; and on the 10th the regiment quitted its quarters and proceeded to Rosendael, where the officers and soldiers received information of the death of King William III. and of the accession of Queen Anne.

The sudden decease of King William, on the 8th March, 1702, did not retard the breaking out of hostilities, as his views were carried into effect by his successor, Queen Anne, who declared war against France and Spain on the 4th May following, and the Earl of Marlborough was appointed to command the forces in Flanders, with the rank of Captain-General. The regiment marched across the country to the Duchy of Cleves, and encamped with the army, under the Earl of Athlone, at Cranenburg, during the siege of Kaysers-werth by the Germans. During the night of the 10th of June the army quitted Cranenburg, to preserve its communication with Nimeguen, in front of which fortress the regiment skirmished with the French on the following morning.

Spanish Netherlands, 1702.

The Earl of Marlborough assembled the army, composed of the troops of several nations, and advanced against the French, who withdrew to avoid a general engagement; and the regiment was afterwards selected to take part in the siege of Venloo, a town in the province of Limburg, on the east side of the River Maese, with a detached fortress beyond the river, against which the British troops carried on their attacks. The regiment took its turn of duty in the trenches, and the grenadier company was engaged in storming the counterscarp of Fort St. Michael on the 18th of September, when the soldiers followed up their first advantage with astonishing intrepidity, and captured the fort.

On this occasion, Lieut.-Colonel Holcroft Blood of the regiment, who was detached from it, whilst performing the duty of principal engineer, highly distinguished himself. (See Biographies of Colonels.)

In a few days after the capture of Fort St. Michael, the besieging army formed to fire a *feu-de-joie* for the taking of Landau by the Germans, when the people and garrison of Venloo, supposing a general attack was about to be made on the town, induced the governor to surrender.

Sir M. Bridges' (17th) Regiment was afterwards employed in the siege of Ruremonde, which fortress was invested towards the end of September, and was forced to surrender, the investment having commenced on the 18th, and ended on the 26th.

Rejoining the main army after the surrender of Ruremonde, the regiment advanced to the city of Liege, and its grenadier company was engaged in the siege of the citadel, which was captured by storm on the 12th of October.

On the afternoon of that date, eight battalions and 1,000 grenadiers advanced to attack the counterscarp, which had been wrecked by four days' pounding with shot. In a letter of the same date to the Minister Nottingham, the Earl of Marlborough writes: "By the extraordinary bravery of the "officers and soldiers, the citadel has been carried by storm, "and, for the honour of His Majesty's subjects, the English "were the first that got upon the breach."* After these conquests the regiment marched back to Holland.

The campaign this year ended on the 31st October.

1703

The Duke of Marlborough arrived in Holland on the 13th March, on return from England, having had conferred on him by his Queen, since the last campaign, the highest title of nobility.

Sir M. Bridges' (17th) Regiment joined in the operations under him, marching towards the end of April in the direction of Maestricht, and it was in position near that city when the French army under Marshal Villeroy and Marshal Boufflers approached, and some cannonading occurred, but the enemy did not hazard a general engagement.

It also took part in the movements which occasioned the French commanders to make a sudden retreat from their position at Tongres, and to take post behind their fortified lines, where the English General was desirous of attacking them, but was prevented by the Dutch generals and field-deputies. The services of the regiment were afterwards connected with the siege of Huy, a strong fortress on the River Maese, above the city of Liege, which was captured in fourteen days.

* "Military Expeditions," 1702-1707, P.R.O.

On the 16th August, Count Noyelles, with part of the Grand Army (which the Confederate army was now termed), arrived before Huy. Notwithstanding the fatigues of a most difficult march, the trenches were opened, and by the 21st the batteries began to throw bombs into the enemy's works. By the following evening the two forts were carried by assault and destroyed, the French retiring into the castle, on which the whole force of the besiegers was now turned. By the 25th preparations were made for a general assault, which, however, was followed by an unconditional surrender. During the fourteen days of the siege the loss of the allies was only 60 killed and wounded, that of the besieged being 200 killed and wounded.*

Another proposal to attack the French lines having been objected to by the Dutch, the regiment was employed in covering the siege of Limburg, a city of the Spanish Netherlands.

The Prince of Hesse was entrusted with all the details of the siege. The investment began on the 10th September, and on the 28th the garrison surrendered, the loss of the besieged having been about 60 killed and wounded, and that of the besiegers not above 100.†

On the 26th August, Lieut.-Colonel Blood was promoted Colonel of the regiment, in succession to Colonel Sir M. Bridges, appointed Governor of Londonderry.

During the summer of this year, Archduke Charles of Austria was acknowledged as King of Spain by England, Holland, and several other states of Europe; and in the autumn of 1703 Portugal joined the Grand Alliance. The coalition against France and Spain arranged that 4,000 Dutch and 6,000 British troops should be sent to Portugal to assist the King of that country and his army in placing Archduke Charles of Austria on the throne of Spain by force of arms.‡

An augmentation of our forces being necessary, the Duke of Marlborough was appealed to as to what regiments he could send from Holland. His reply, dated Borchloen, July 29th, 1703, recommended that, four regiments to be sent, should be Portmore's (2nd Queen's), Stanhope's (11th), Stewart's

* Millner's Sieges.
† R. Cannon. Millner's Sieges, p. 73.
‡ Dalton.

(9th), and Sir Matthew Bridges' (17th), "all four old regiments, "and I think, very good ones; the last two are the strongest "we have." * Accordingly, a Royal Warrant, dated August 20th, was issued by Queen Anne, that these regiments should be brought to England for service in Portugal, to be made up to 13 companies of three sergeants, three corporals, one drummer, and 56 privates in each, including servants.

To England, Nov. 1703.

The regiment, accordingly, embarked from Holland on the 20th November and sailed to Portsmouth, where it was detained by contrary winds.

1704

Portugal, 1704.

It put to sea in January, but, encountering a severe storm, was driven back to port, and several ships of the fleet were much damaged. The voyage was afterwards resumed, and the regiment forming a part of 6,000 troops on board the fleet (under the Duke of Schomberg, who had been appointed Commander-in-Chief to the Queen's forces to be employed in Portugal), arrived at Lisbon in the early part of March, and landed on the 15th of that month. The King of Portugal being afraid to entrust the protection of his frontier towns to his own troops, the British regiments were placed in garrison.

Great tardiness in preparations for opening the campaign was manifested by the Portuguese authorities, and the Duke of Berwick (one of the ablest French generals, at the head of a large and well equipped army) attacked the frontiers of Portugal with the combined French and Spanish armies before the allies were prepared to take the field. The regiment was called from garrison to take part in attempting to arrest the progress of the enemy; it was employed in the Alemtejo district, and in July was encamped with the Queen's, Earl of Barrymore's (13th), Duncanson's (33rd), Lord Mountjoy's Regiment, and Brudenell's Artillery near Estremos—a town situate in an agreeable tract on the Tarra; towards the end of July they marched into cantonments in the town.†

* Duke of Marlboro's Despatch.
† Col. Davis's History, "2nd Queen's."

The Duke of Schomberg was at this period recalled, and the Earl of Galway appointed Commander-in-Chief in Portugal, arriving at Lisbon on the 11th August.

In the autumn, the allied army was enabled to act on the offensive, and Colonel Blood's (17th) was one of the regiments which penetrated Spain to the vicinity of Ciudad Rodrigo; but the enemy was found so advantageously posted, beyond the Agueda, that the Portuguese generals objected to attempt the passage of the river, and the army returned to Portugal, where the regiment passed the winter.

1705

The forces of the allies were as follows:—English, one regiment of horse; five of foot, including the Queen's, with Colonel Blood's (17th), Stewart's (9th), Duncanson's (33rd), with artillery, one train of 5-pounders. Other forces: Dutch, 2,300; Portuguese, 12,000. Colonel Blood's Regiment again proceeded to Estremos, in the Alemtejo, in April, 1705, and, with the allies under Lord Galway, it was engaged in the siege of Valencia de Alcantara, which place was captured by storm on the 8th of May. The regiment was also employed at the siege and capture of Albuquerque.

Spain, 1705.

The place was situated on a hill, surmounted by a castle, which was accessible on one side only. The garrison consisted of 800 Spaniards. Lord Galway commenced his attack on the 16th May by storming the suburbs of the town, and, as the artillery fire could make little impression on the walls, which were of very solid and strong masonry, mining was commenced. On the 20th, a breach having been made in the wall of a church near the ramparts, a party was sent to gain a position in it, when the garrison, being completely surprised and finding further resistance useless, surrendered.*

On the 4th June, the march of the allies from Albuquerque towards Badajoz commenced at 4 a.m., and, on the 6th, a council of war decided to at once begin the siege, for which preparations were in progress, when, on the 13th, orders arrived from Lisbon to separate into summer quarters. In obeying these orders the English force of the allies crossed the

* Parnell, p. 101.

Caya River and encamped along the Andalusian frontier, Colonel Blood's (17th) taking up quarters at Moura, near the Guadiana River. On the 30th September, Lord Galway with the English troops crossed the frontier, and the regiment was engaged in the second siege of Badajoz, which was not more successful than the first. It was being carried on with great vigour until Lord Galway lost, by a cannon ball, his right hand, which caused amputation at the arm. The command then devolved on the Dutch and Portuguese commanders, who were out-manœuvred by the French throwing 1,000 troops into Badajoz. The siege was raised and, by the 17th October, the allies had begun their retreat, which was effected without any loss, and on arrival at Elvas the army separated and went into winter quarters.

1706

The forces that were now to be matched against each other were as follows: The allies had with them 19,000 troops. The English force was 200 Harvey's Horse and 2,000 foot, composed of Lord Portmore's (2nd), Colonel Stewart's (9th), Blood's (17th), Wade's (33rd), and Brudenell's. The English artillery was represented by ten field-pieces. The Dutch had four squadrons and 2,000 foot, and the Portuguese had 3,600 horse and 11,100 foot, their artillery consisting of eight light and 24 heavy guns.* The Duke of Berwick's force consisted altogether of 47 squadrons of horse, and 27 battalions of foot; in all 15,300 men, all Spanish troops. After passing the winter in cantonments on the borders of Portugal, the Earl of Galway again took the field on the 30th March, 1706, when the allies marched to Salvador, and it was decided to advance to the frontier and besiege the fortress of Alcantara, a fortified town situated on a rock near the River Tagus, in Spanish Estremadura. A sharp cavalry encounter took place at Brocas on the 8th April, in which the French were repulsed with a loss of 100 killed and prisoners, and on the 9th Alcantara was reached. The troops commenced to entrench themselves the first night, throughout which the French kept up a brisk fire, and next day made a vigorous

* MSS. British Army, R.U.S.I.

sally, which was repulsed with such loss that, on the 14th, they surrendered the town, leaving the allies in possession. Colonel Blood's (17th), and Colonel Wade's (33rd) Regiments gained great distinction in this siege for their gallantry at the assault of the outlying convent of St. Francis near the town, which the enemy had converted into an outlying fort. The two regiments had 50 officers and men killed and wounded, and, amongst the latter, were three captains of Colonel Blood's Regiment.* After the capture two Portuguese regiments were directed to garrison the convent.

The plunder included, besides 70 guns and mortars, 5,000 muskets, 22,000 pounds of corn, 200 pipes of wine, 150 pipes of oil, and 12,000 suits of men's clothing.† The capture of Alcantara was of great value to the allies, as in addition to their capturing stores and munitions of war, it considerably weakened the fighting strength of the enemy.

From Alcantara the army advanced to the vicinity of Placencia, and afterwards drove the enemy from his position on the banks of the Tietar—sending forward a detachment to destroy the bridge of Almaraz; but, subsequently changing its route, proceeded to the province of Leon, and Colonel Blood's (17th) Regiment was employed in the siege of Ciudad Rodrigo, which fortress surrendered on the 26th of May. The town, which was surrounded by walls of little strength, was soon breached by the guns of the allies, and a storming party prepared.

Finding resistance useless, the governor capitulated.

On the 3rd of June the army, with provisions for 24 days, commenced its march from Ciudad Rodrigo for the capital of Spain, proceeding by Salamanca, through the Guadarrama Mountains; and, arriving at Madrid on the 24th and 27th of June, encamped in the vicinity of that city, where, on the 2nd July, Archduke Charles of Austria was proclaimed King of Spain with the usual solemnities. This tide of success was changed by the delay of King Charles to come to Madrid from Barcelona, which fortress had been captured by the Earl of Peterborough in the preceding year. This delay occasioned his friends to be discouraged; the partisans of

* Col. Davis's History of the "Queen's."
† Parnell, p. 174.

King Philip took up arms ; and, the whole country responding to his proclamation, numerous bodies of French and Spanish troops joined the army under the Duke of Berwick, increasing his force to twice that of the allies. King Charles and the Earl of Peterborough joined Lord Galway on the 6th August with 3,000 men, but the Earl, being unable to agree with the other commanders, soon left. Finally, Galway, cut off from his base in Portugal, marched a month later (September 17th) for Valencia, where Colonel Blood's (17th) Regiment was stationed during the winter.

1707

On the 1st January, this year, Colonel Blood was promoted Major-General. Early in April, 1707, the regiment joined the allied army under the Marquis das Minas and the Earl of Galway, which now numbered 15,500, and after taking part in several operations, advanced, on the 25th April, to attack the French and Spanish troops, under the Duke of Berwick, at Almanza, whose total strength numbered about 25,000.

The effective strength of the English regiments with Galway, at the battle of Almanza (from the weekly return dated April 22nd, 1707), shows a total of 8,900, as follows :—

HORSE.

Harvey's Horse, now 2nd Dragoon Guards	.. ..	227
Carpenter's Dragoons, now 3rd Hussars	.. ..	292
Killigrew's ,, now 8th ,,	.. ..	51
Pearce's ,, disbanded	.. .. ..	273
Peterborough's ,, ,,	.. .. ..	303
Guiscard's ,, ,,	.. .. ..	228

FOOT.

Foot Guards	.. ..	400
Portmore's, now 2nd Queen's	.. ..	462
Southwell's, ,, 6th Foot	.. ..	505
Stewart's, ,, 9th ,,	.. ..	467
Hill's, ,, 11th ,,	.. ..	472
Blood's, ,, 17th ,,	.. ..	461
Mordaunt's, ,, 28th ,,	.. ..	532

Wade's,	now 33rd Foot	.. ..	458
George's,	" 35th "	.. ..	616
Allnutt's,	" 36th "	.. ..	412
Mountjoy's,	disbanded	.. ..	508
Macartney's,	"	.. ..	494
Breton's,	"	.. ..	248
John Caulfield's	"	.. ..	470
Lord Mark Kerr's	"	.. ..	419
Count Nassau's	"	.. ..	422
		Total ..	8,900

On the night of the 24th April the allies rested at Caudete, and at daybreak, on the 25th, marched in four columns towards Almanza, a distance of about eight miles.

The following account of the battle of Almanza, fought on Easter Sunday, is mainly taken from Lord Galway's despatch, in the "London Gazette," dated June 2nd, 1707:—

About 3 p.m. the Earl of Galway posted himself at the head of the British dragoons and marched to begin the battle with the enemy's right wing of horse, the Portuguese being ordered to take the charge, but not before the English and Dutch were actually engaged. The enemy, observing that the Portuguese cavalry of our right did not advance with our left wing, ordered some squadrons to attack them, which they did with such success that the Portuguese were utterly broken, by which their infantry were immediately surrounded, and most of them killed or taken prisoners. Berwick then finding that he endeavoured to no purpose to break our left with horse only, sent for nine battalions, mostly French, to oppose our brigade of foot, which consisted of Colonel Southwell's (6th), Blood's (17th), Wade's (33rd), and Mountjoy's Regiments, which were reinforced by Lieut.-General Stewart's (9th) Regiment from the rear line. At the same time they brought up several fresh squadrons to charge our left wing of horse, which had suffered very much. Our troops, in this condition, were not able to sustain their charge, and gave way, at which time the nine French battalions charged the English brigade of foot in front and flank, and entirely broke them.

A most vigorous charge was now made by Harvey's Horse (2nd Dragoon Guards) on two French battalions that had attacked us in flank, by which the regiment broke through them, and made them beg for quarter before their own cavalry could come to their assistance. The fight still raged in the centre, but the flanks being defeated the enemy surrounded the centre, and made great slaughter, whereupon General Shrimpton and several officers assembled the stragglers of the English regiments, and, joined by some of the Dutch and Portuguese, who had been similarly rallied, formed a body of nearly 2,000 men, who retreated to the hills of Caudete, the enemy's cavalry still pursuing, though often repulsed by the fire of our foot.

The men being exhausted from the fatigues of the day, and in want of both ammunition and provisions, were unable to march further, and the next morning, being surrounded by two lines of foot, the commanding officers agreed to the same capitulation that had been granted to the French at Blenheim, and surrendered themselves prisoners. The English regiments which capitulated were Portmore's (2nd), Hill's (11th), George's (35th), and Macartney's and Breton's.*

The officers and soldiers of Major-General Blood's (17th) who escaped from the field, joined the Earl of Galway, who had retreated with the cavalry that remained (about 3,500) to Alcira, on the River Xucar, and the approach to the town being by almost inaccessible mountains, his lordship halted there a few days to reorganise the army.

The regiment had the following casualties :—

Killed : Lieut.-Colonel Daniel Woollett, Lieut.-Colonel George Withers, and Major Anthony Leech. Wounded and taken prisoners : Captains Fitzgerald and James Fonsubran, Lieutenants John Rivason, Wm. Ingram, and Edmond Blood ; Ensigns Carlow, John Dumaresq, and Wm. Bruce. Prisoners : Captains Dudley Cosby and Loftus Cosby, Lieutenants Edw. Martin, John Brown, Robert Mountford, Benjamin Brooks, Edw. Tyrrell and Ensign Bland, and a great number of non-commissioned officers and men killed and wounded.

The author of " The Annals of Queen Anne " observes : " Had the Portuguese bravely seconded the English and Dutch, " who, with unparalleled resolution and undauntedness, broke " the enemy's centre, it is the opinion of many that victory

* Parnell, p. 220.

"would have inclined to the confederate side, or, at least, "that the latter might have made an honourable retreat, "and, considering the vast disproportion of the forces, have "gained the glory of the day."

Out of the English stragglers from Almanza, Galway organised five new battalions, and with them resuscitated five of the oldest regiments that had been broken up in the battle, viz.: Portmore's (Queen's), Southwell's (6th), Stewart's (9th), Hill's (11th), and Blood's (17th).

The five were then reduced to four, the Queen's being the one that was taken.*

Major-General Blood died at Brussels on the 19th August, 1707, whilst serving as Colonel of the English artillery train under the Duke of Marlborough, and the Duke, in a despatch announcing his death, refers to him as being "much lamented "for his bravery and experience."†

Queen Anne conferred the colonelcy of the regiment on Colonel Joseph Wightman, who had been promoted Brigadier-General on the 1st January, 1707.

Lord Galway, leaving garrisons of foot at Alcira and Xativa, marched to the further side of the Ebro near Tortosa, on the banks of which the regiment was encamped for some time, and was afterwards employed in operations for the protection of Catalonia. It was joined by men from command and sick absent, also by several who escaped as prisoners of war (many of whom had been forced into the service of the French King, but deserted), and it mustered 266 officers and men, its strength on going into action having been 461.

The first complete code of "Clothing Regulations" was issued during the reign of Queen Anne, dated 14th January, 1707, and states that:

"The sole responsibility for the pay and equipment of "a regiment rests with the Colonel, who is held responsible "in his fortune, and in his character, for the supplies of his "regiment."

By Warrant of Queen Anne, in 1707, a Board of General Officers was instituted, and the duty of this Board was to select, seal, and issue patterns for the clothing of each regiment by

* Col. Davis's History, "2nd Queen's."
† "London Gazette," Sept. 1st, 1707.

the colonel. The supplies furnished by the colonel, when received at the regiment, were inspected by a Regimental Board, and compared with the patterns sealed by the Board of General Officers. Any complaints made by the regiment were investigated by the Board, whose decision was final.*

The regiment was encamped for some time on the river Francoli, between Monblanco and Tarragona, and afterwards at Constantina.

As a result of the heavy casualties among officers of the regiment at Almanza, there were no less than thirty new officers appointed to it by commissions, dated at Kensington, in the following month of March.†

* Old Pimlico Book.
† "George the First's Army," Dalton, p. 49.

Major-General the Honourable Robert Monckton.

1761.

CHAPTER III.

GREAT BRITAIN, MINORCA, IRELAND, AND NOVA SCOTIA.

1708–1757.

1708

IN a letter from the army in Spain, dated Tarragona, April 23rd, 1708, it was stated: "We cannot yet give any certain "account of the number of our forces, but those we have are "the finest in the world; such are the regiments of Southwell "(6th), commanded by Lieut.-Colonel Hunt, that of Blood "(17th), commanded by Lieut.-Colonel Bourguet, and that "of Mordaunt (28th), commanded by Colonel Dalziel."*

Early this year the regiment received orders to return to England, where it arrived on or about the 12th April, and commenced, by beat of drum, to recruit its numbers, which soon became largely increased by officers' recruiting parties at Stafford, Norwich, and Exeter, as shown by the marching orders at that period.

On arrival, Brigadier-General Wightman was sent for, to Whitehall, for the Secretary at War to "settle with him "the headquarters of his regiment." †

The destination decided on was York, where, however, it did not stay long, as a Route dated 8th May directed eight companies of Brigadier Wightman's Regiment to march from York to Berwick, and five from York to Hull, and by the end of the year the whole regiment was concentrated at Hull.

* This letter was published in the "State of Europe," for June, 1708, but the writer was evidently neither aware of Brigadier-General Wightman's appointment to command the regiment, in succession to Major-General Blood, deceased, nor of General Wightman's Regiment having embarked for England early in 1708.

† W.O. 4, Book 24, Public Records Office, London.

1708

In a Royal Warrant, dated Windsor, 12th July, Queen Anne decided to abolish all brevet rank in the army.

A Royal Warrant, dated August 23rd, laid down the following "annual bounties," to be paid to the several widows of officers who had been killed or died in Spain, Portugal, or the West Indies, viz. :—

The widow of a	Colonel	£50	per annum.
,,	Lieut.-Colonel ..	40	,,
,,	Major	30	,,
,,	Captain	26	,,
,,	Lieutenant ..	20	,,
,,	Ensign	16	,,
,,	Cornet	16	,,
,,	Adjutant	16	,,
,,	Quarter-Master ..	16	,,
,,	Chirurgeon (Surgeon)	16	,,
,,	Chaplain	16	,,

1709

Scotland. 1709.

In February the regiment received orders to march, in three divisions, from Hull to Liverpool, "when relieved by "three regiments from Ostend." A Route, dated 7th April, directed it to march from Liverpool to Edinburgh, where it was joined by a recruiting party of sergeants, corporals, and drums from Bowe.

1710

On the 1st January, this year, Brigadier-General Wightman was promoted Major-General, and the headquarters were stationed at Leith, with a detachment of four companies at Musselburgh.

1711

Queen Anne issued a Royal Warrant on the 1st May, regulating the conditions for obtaining commissions in the army, and for retirements by the sale of them, and also directed that no more brevets were to be granted on any pretence whatever.

1712

An order, dated 30th July, directed the reduction of the regiment to 12 companies, each consisting of one captain, one lieutenant, one ensign, two sergeants, three corporals, two drummers, and 50 privates, including officers' servants. Any non-commissioned officers and soldiers in excess were to be disbanded, and allowed 14 days subsistence to carry them home. They were also permitted to take with them their clothes, belts and knapsacks, and to be paid 3s. each for their swords, which, with their arms, were to be returned into the Ordnance Stores. The sergeants were allowed to keep their swords. The commissioned officers disbanded were to be put on half-pay.*

An order of the same date directed Major-General Wightman's (17th) to march to Fort William (Inverness) to relieve Colonel Windress's Regiment, ordered to Ireland from the 24th August.†

1713

The Treaty of Utrecht was signed on the 13th March, when the regiment was placed on the peace establishment.

By a Royal Warrant of Queen Anne, dated 23rd April, the following order of precedence was laid down for the first 18 old British infantry regiments, when the 17th came in the order which has since been allotted to it :—

1. Royal.
2. Colonel Kirke's.
3. Colonel Selwyn's.
4. Our Own Regiment (Lieut.-General Seymour).
5. Major-General Pearce's.
6. Colonel Hamilton's.
7. Royal Fusiliers.
8. Our Own Regiment (Lieut.-General Webb).
9. Lieut.-General Stuart's.
10. Lord North and Grey's.
11. Major-General Hill's.
12. Colonel Phillips's.

* W.O. 26, Book 14, p. 3.
† W.O. 26, Book 14, p. 7.

13. Earl of Barrymore's.
14. Lieut.-General Tidcombe's.
15. Earl of Hartford's.
16. Late Brigadier Darell's.
17. Major-General Wightman's.
18. Royal Regiment of Ireland.*

1714

Ireland, 1714.

The regiment had returned to Leith by the 20th March this year, and, on the 2nd April, orders were received to embark for Ireland. It accordingly marched to Port Patrick for embarkation, and arriving on the 24th April, landed at Donaghadee.

On the 14th June, the following moves were ordered: Seven companies to Newtown, Belfast and Carrickfergus; one to Newry, one to Carlingford, and one to Dundalk, Garlandstown, Dunleese and Drogheda, and several interchanges of quarters between companies took place up to the spring of the following year.

1715

Scotland, 1715.

On the 23rd March, the regiment was ordered to march from its present quarters to Dublin, where it remained until 29th September, when it was withdrawn from Ireland; and on the breaking out of the rebellion of the Earl of Mar in favour of the Pretender, it joined the troops encamped at Stirling under the Duke of Argyll, and was in brigade under its colonel, Major-General Wightman.

When the rebel army advanced with the view of penetrating southwards, the King's troops quitted the camp at Stirling and proceeded to the vicinity of Dunblain, and on the 13th of November an engagement took place on Sheriff Muir, which is related as follows in the Duke of Argyll's despatch, "London Gazette," dated November 19th, 1715:—

BATTLE OF SHERIFFMUIR.

On the 12th November the Duke of Argyll, finding that the rebels had come to Auchterarder, with their artillery, baggage, and stores, considered it necessary either to engage

* W.O. 26, Book 14.

them on the grounds near Dunblain, or to decamp and await their coming to the Head of Forth. He chose the former; and on ascertaining that they intended to encamp that night at Dunblain, he marched in the forenoon, and encamped with his left there and his right towards the Sheriff-Moor, the enemy encamping within two miles of Dunblain. The Duke, with Major-General Evans, commanded on our right, Major-General Wightman in the centre, and Major-General Whetham on the left. The next morning (13th), perceiving that the rebels purposed making a flank attack, His Grace ordered his troops to stretch to the right in the following order: Three squadrons of dragoons upon the flanks of his front line, and six battalions of foot in the centre. The six battalions in the centre consisted of Clayton's (14th), Montague's (11th), Morrison's (8th), Shannon's (25th), Wightman's (17th), and Forfar's (3rd Buffs).

The second line was composed of two battalions, Orrery's (21st) and Egerton's (36th), with a squadron on each flank, in rear of the squadrons in the front line.

His Grace, finding that the rebels were not quite formed, gave immediate orders to charge both their horse and foot. They received us very briskly, but, after some resistance, they were broken through, and pursued for over two miles, by five squadrons of dragoons, the squadron of volunteers, and five battalions of foot. When we came near river Allan, from the vast number of rebels we drove before us, we concluded it was an entire rout. Major-General Wightman, who commanded the five battalions of foot, sent to acquaint the Duke that he could not discover what had become of our troops on the left, and that a considerable number of the rebel horse and foot were behind us.

Thereupon His Grace, marching his troops towards the hill on which the rebels had posted themselves, extended his right towards Dunblain to give his left an opportunity of joining him. Not finding our left come up, the Duke marched to the position he had taken up in the morning. When it was dark the rebels moved to Ardoch, and about an hour later, our troops who had become separated from the Duke of Argyll, joined His Grace. Our dragoons on the left, in the beginning of the action charged some of the rebel horse on the right and carried off a standard, but the rebels pressed so hard on our battalions on the left, that the latter were disordered and obliged to fall in among the horse. By this means they cut off communication between our left and the main body, and, being informed that a body of them were endeavouring to get to Stirling, the troops of our left retired beyond Dunblain to obtain possession of the passes leading to it.

The number of the rebels killed is reckoned as about 200, and we have captured 14 colours and standards, 4 guns, tumbrels with ammunition, and all their bread waggons.

The courage of the British troops was never keener than on this occasion ; though the rebels were three times their number, they attacked and pursued them with all the resolution imaginable. The conduct and bravery of the generals and other officers contributed much to this success. On the night of the 13th, that body of the rebels who had made a stand after their left wing was defeated, retired beyond Auchterarder, and, as some say, to Perth.

Upon this, the Duke of Argyll, who had purposed attacking them on the morning of the 14th, finding all the forage about Dunblain consumed and that there was a difficulty in getting provisions for His Majesty's troops, returned that evening to Stirling and brought with him the plunder he had taken. The rebels were prevented marching southwards, and they did not hazard another engagement, which proved the advantage gained over them.

A Return of the casualties on this occasion shows :

Wightman's (17th), Ensign Mark wounded ; two grenadiers, and two or three men killed.*

A Commemorative Medal† for "Sheriffmuir or Dunblain, 13 Nov. 1715," was struck, of the following description :

Obverse : Bust of George I, r., laureate, hair long, in figured armour, and mantle fastened with brooch on the shoulder. Legend, Georgius D : G : Mag : Br : Fr : Et : Hib : Rex. F : D. Below I. C.

Reverse : Victory with sword and palm branch, rapidly pursuing a body of fleeing cavalry. Legend, Perjurii ultrix. (The Avenger of perjury.) Ex. A.D. DVNBLAINVM. 13 Nov. 1715. (At Dunblain, 13 Nov. 1715.) (See Plate 5.)‡

* Robert Patten's History of the Rebellion.
† By the courtesy of C. Dalton, Esq., F.R.G.S.
‡ "Medallic Illustrations."

1716

Additional forces having joined the Royal army, the Duke of Argyll advanced in January, over ice and through snow, towards Perth, when the Pretender retreated, and soon afterwards fled, with the leaders of the rebellion, to France. General Wightman's (17th) Regiment pursued the insurgents some distance, and was afterwards stationed at Perth.

1717

A War Office order, dated 8th November, authorised the following establishment of the regiment (10 companies), viz.: one captain, one lieutenant, one ensign, two sergeants, three corporals, one drummer, and 35 privates per company.

1718

A Route, dated 13th March, directed Major-General Wightman's (17th) to march to Berwick and thence to Newcastle, the headquarters and six companies returning to Berwick in November, with detachments at Newcastle, Gateshead, and Tynemouth.

1719-1721

King George I. assented to new regulations, this year, for the sale of officers' commissions.*

On the 3rd March, 1719, the regiment was ordered to march to Hull, and on the 5th May the headquarters and five companies were transferred to Newcastle and Gateshead.

On the 21st May, the headquarter detachment, at Newcastle, was employed with six companies of the "Queen's" in quelling a riot of keelmen.

On the 2nd July, it marched in three divisions from Newcastle to Edinburgh, and had returned to Newcastle prior to July, 1721, as on the 6th of that month orders were issued to march from Newcastle to Chester, whence it proceeded to Waterford, and was taken on the strength of the Irish establishment from the 30th September.

Ireland, 1721.

1722

A marching order, dated 17th August, directed the regiment to embark with five others ordered to England.

England, Aug., 1722.

* W.O. 26, Vol. XVI.

On arrival it remained with one regiment at Bristol, whilst two proceeded to Chester and two to Wells.

On the 25th September, Major-General Wightman died, and King George I. conferred the colonelcy of the regiment on Brigadier-General Thomas Ferrers, from the 39th Regiment, and this officer dying on the 26th October, he was succeeded, on the 7th November, by Colonel James Tyrrell, who had commanded one of the regiments of dragoons disbanded in 1718.

A War Office letter, dated, 22nd November, directed the return to Bristol of the four regiments that had left it in August, and all the colonels of the six regiments from Ireland were to hire shipping, convenient to transport their regiments to Cork and Kinsale, and also, to furnish them with provisions for the passage as cheaply as possible. Colonel Tyrrell's (17th) sailed in three detachments, the last of which reached Kinsale on the 31st December, where it remained.

Ireland, Dec., 1722.

1723

An official State of the army in Ireland, for October, shows the regiment mustering 50 non-commissioned officers and 380 privates,* and it continued at Kinsale throughout the following year.

1725

On the 2nd April, orders were issued for the regiment to march to Cork, where, on arrival, it was held in readiness to embark for Minorca.† It embarked on the 11th July, (strength, 29 officers, 1 surgeon's mate, 20 sergeants, 30 corporals, 10 drummers, 319 sentinels), and, from the 14th August, was placed on the establishment of Minorca (the second of the Balearic Islands in the Mediterranean, near the coast of Spain), which had been captured by the British in 1708, and was ceded to Great Britain at the Peace of Utrecht in 1713. It is about thirty miles in length and twelve in breadth, and is chiefly valuable for the excellent harbour of Port Mahon, with deep water, and capable of sheltering in former days all the fleets of Europe. At the entrance to it is the castle of San Felipe.

Minorca, 1725.

* Irish "Martial Affairs," Vol. XVII.
† Irish "Martial Affairs," Vol. XVIII.

The people of the island were well housed in solid stone buildings, the farmhouses being generally of two stories, with the granary under the roof. The farmers have to contend against frequent and violent gales, a very stony and shallow soil, and scarcity of water. They are very laborious, and work under a system of partnership.*

1727

Since the siege of Gibraltar in 1705, it had remained unmolested until the latter end of 1726, when the Spaniards, who had kept a watchful eye on the garrison, assembled an army in the neighbourhood of Algeciras, which led to hostilities.

On the 22nd February, the Count de las Torres, commanding the Spanish force, opened fire on the garrison with 17 guns, besides mortars. Throughout the siege the Spaniards were much harassed by the ships of our fleet, under Sir Charles Wager and Admiral Hopson. A detachment of Colonel Tyrrell's (17th) Regiment proceeded early in the year with a mixed force from Minorca to assist in the defence of the fortress, but there is no record of any particular part it took. Reinforcements of British troops were frequently arriving, and amongst them, a detachment of 500 men from Minorca in April, under Colonel Cosby. By the 26th the Spanish batteries amounted to 60 guns in addition to mortars.

On the 3rd May, information was received of an intended assault on the garrison, and precautions were accordingly taken. The firing continued until the 12th, when at 10 p.m., letters were delivered to Lord Portmore, from the Dutch minister at the Court of Madrid, with a copy of the preliminaries of a general peace, whereupon all hostilities ceased, and the detachment of troops from Minorca returned to that island.

The garrison lost about 300 killed and wounded, and 17 guns and 30 mortars had burst during the siege. The enemy's casualties could not be ascertained, but their loss was computed at nearly 3,000.†

Colonel Tyrrell was promoted this year to the rank of Brigadier-General.

* Sir Clements Markham, K.C.B., "Story of Majorca and Minorca," pp. 263–9.

† History, by Capt. John Drinkwater, pp. 18 to 22.

1729

The size of a horse for a trooper, at this period, was described as " a strong well-bodied horse from fifteen hands " and an inch, to two inches, and not exceeding."

1732

A Return of the troops in Minorca, dated 25th March, shows the garrison to consist of four British infantry regiments, including Brigadier-General James Tyrrell's (17th), which mustered 19 officers, 4 staff, 30 sergeants, 30 corporals, 20 drummers, 441 private men, and 59 wanting.*

1735

Brigadier-General Tyrrell was promoted on the 7th November this year to Major-General.

1739

A Royal Warrant, dated 12th June, directed the regiment to be increased by 100 men (10 to each company), and a Secretary at War's letter, dated 27th September, notified that each company was to consist of three sergeants, three corporals, two drummers, and 70 privates.

Major-General Tyrrell was promoted this year to Lieut.-General.

1740

On the 7th May, the following scheme was drawn out for soldiers serving as marines on board His Majesty's ships :—

On board a ship of 90 to 100 guns, a complement of one captain, one lieutenant, and 100 men ; in a ship of 80 guns, two lieutenants and 80 men ; in one of 70 guns, the same ; in one of 50 to 60 guns, one lieutenant and 60 men ; in one of 40 guns, one lieutenant and 50 men ; in one of 20 guns, one lieutenant and 30 men. In sloops, 15 men supernumerary to their complement of 150 men.†

In John Armstrong's " History of Minorca," a letter, dated 19th June, says : " We have now in the island five old " regiments of foot, viz., Brigadier Read's (9th) ; Lieut.-

* W.O. 1, Book 294. † W.O. 26, Book 19.

"General Tyrrell's (17th); the Royal Regiment of Foot of "Ireland (18th), commanded by Major-General Armstrong; "Brigadier Paget's (22nd); Major-General Anstruther's (26th). "Of these, only one-third can be put on duty at once, and we "have a vast extent of works for 800 men to defend."

1742

On the 1st August Lieut.-General Tyrrell died, and on the 31st August the colonelcy was conferred on Colonel John Wynyard, from the 4th Marines, who had previously held the commission of lieut.-colonel in the 17th Regiment upwards of 20 years.

1747

An official order appears this year that "no officer under "the degree of a brigadier shall ever appear, either in quarters "or in camp, whether on duty or off, in any other coat than "his regimentals or uniform, either old or new. No officer "under the degree of a brigadier shall have either chariot or "chaise."

1748

At the termination of the War of the Austrian Succession, when peace was concluded at Aix-la-Chapelle, the regiment was relieved from duty at Minorca, and taken on the strength of the Irish establishment from the 25th December,* though it did not arrive home until the autumn of the following year.

1749

Ireland 1749.

A letter, dated Dublin Barrack Office, 14th August, states that Routes were forwarded on that date for the four regiments expected to land at Cork or Kinsale from Minorca, of which Major-General Wynyard's (17th) was one. A muster roll taken on landing shows that the regiment disembarked on the 5th October, when it marched from Cork and Kinsale to Galway, and later in the month three companies were detached to Ballinrobe, three to Castlebar, and two distributed between Headford, Ballinrobe, Castlebar, and Foxford Barracks.†

* Secretary at War's letter, dated 18th Nov., 1748.
† Irish Routes, Bundle 14.

A Route, dated 23rd December, directed the headquarters and five companies to move from Galway town to Galway New Barracks.

1750

On the 2nd June, the regiment received orders to march from Galway to Cork to be reviewed (the term used at that period for "inspected").

1751

A Route, dated 6th June, directed it to march from Cork to Kinsale to be reviewed; and thence, headquarters and four companies to Bandon, two companies to Kinsale town, two to Charlesfort, and two to Clonakilty.

On the 1st July a Royal Warrant was issued, assigning, for the first time, regimental numbers to cavalry and infantry regiments, and also regulating the clothing, standards, and colours of the several regiments of the army, which (as far as the 17th is concerned) is referred to in Chapter XV.

1752

Lieut.-General Wynyard died this year, and King George II. nominated Brigadier-General Edward Richbell to the colonelcy of the regiment from the 14th March.

On the 14th April, the detached companies received orders to rejoin headquarters at Bandon, for the regiment to be reviewed, after which it marched to Cork.

1753

A Route, dated 14th May, directed the regiment, after being reviewed, to march from Cork, and by the 23rd May it was distributed as follows: Three companies at Rosscastle, two at Dingle, two at Castle Island, two at Millstreet, and one at Macroom.

1754

On the 30th March, two companies at Rosscastle were transferred to Nedden, and a Route, dated, 24th April, directed the regiment to march from its present quarters to Dublin.

1755

On the 16th May, the regiment received orders to march from Dublin to Galway in three divisions, and on the 29th of the month, one and a half companies proceeded to Athenry and two and a half companies to Headfort, with detachments at seven out-stations. The Army List of this year, compiled to the month of August, shows the headquarters at Foxford, six companies at Galway, and two at Ballinrobe.

1756

By a Royal Warrant, dated 11th March, the companies of the regiment were each augmented to a total of three sergeants, three corporals, two drummers and 70 privates per company, to take effect from the 31st January.

A Route, dated 20th July, directed the regiment to march from Galway to Kilkenny, detaching two companies to Limerick, which rejoined headquarters on the 14th August.

On the 20th September it was ordered to be reviewed by Lord Rothes.

1757

On the 3rd February an order was issued for the regiment to march to Cork.

Nova Scotia, 1757.

Brigadier-General Richbell died on the 24th February, and was succeeded on the following day by Colonel John Forbes, from the lieut.-colonelcy of the Scots Greys.

Another war having now commenced with France, the regiment embarked at Cork on the 5th May, and sailed to Halifax, Nova Scotia, where it arrived in the early part of July, in order to take part in an attack on the French possessions in Canada: but the expedition was deferred until the following year, and the regiment remained in Nova Scotia during the winter.

In a Return of troops under the Earl of Loudoun's immediate command, dated 24th July, 1757, the 17th Regiment is shown as 700 effective rank and file, with 30 wanting to complete.

CHAPTER IV.

NOVA SCOTIA, SIEGE OF LOUISBURG, NORTH AMERICA, CANADA, MARTINIQUE, HAVANNAH, NORTH AMERICA, AND BACK SETTLEMENTS OF CANADA.

1758–1766.

1758

Nova Scotia, 1758.

WHEN Canada was restored to France in 1632, the island of Cape Breton, in the Gulf of St. Lawrence, was included in the cession, and remained till 1745 in possession of the French, who fortified the town and capacious harbour of Louisburg at great expense.

In April, 1745, an expedition for its capture had been organised by the local authorities at Boston, and approved by the home government, which resulted in its capitulation, and, with it, the whole island of Cape Breton, on June 17th of that year. The island was, however, restored to France in 1748. In 1757, the British sent a large naval and military force against Louisburg, with no good results, so when the expedition to Cape Breton was decided on in the following year, Major-General Amherst (afterwards Lord Amherst) was appointed to the chief command, with Colonels Whitmore, Lawrence and Wolfe as Brigadiers.

In addition to the fleet, the force, consisting of 14 infantry battalions, with a due proportion of artillery, sailed from Halifax on the 28th May, including the 17th Regiment, under Lieut.-Colonel Arthur Morris.

In a Distribution Return of Troops for the Siege of Louisburg, dated 20th February, 1758, the establishment of the regiment is shown as 30 officers, 5 staff, 30 sergeants, 20 drummers, and 690 rank and file.

The order of General Amherst's force in brigade was:

1st Brigade (Whitmore's): 1st Battalion Royals, 22nd, 40th, 48th, and 3rd 60th;

Engraved for Murray's History of the American War.
GENERAL MONTGOMERY.

1775.

2nd Brigade (Lawrence's): 15th, 28th, 45th, 58th, and Fraser's Highlanders;

3rd Brigade (Wolfe's): 17th, 35th, 47th, and 2nd 60th.*

Siege of Louisburg

The whole force arrived off Cape Breton on June 2nd, and anchored in Garbarus Bay, about seven miles from Louisburg, but in consequence of heavy gales, a landing could not be effected until the 8th.

Wolfe's Brigade was to land first, while the other two brigades pulled along the shore to distract attention. As the boats neared the shore, the French, who had reserved their fire, opened a terrific fusilade on them, and though the surf was so great as to make landing nearly impossible, Brigadier Wolfe pushed ashore, and landing on the left of the enemy's works, he formed up quickly, and briskly engaged and routed them. The remainder of the army followed the example without loss of time, landing almost up to their waists in water. The ardour of the troops was not to be conceived nor paralleled; many boats were destroyed and many brave fellows drowned, yet our whole loss on landing did not exceed 110 of all ranks killed, wounded and drowned. The enemy's flight was through the roughest and worst of ground, and Brigadier Wolfe pursued them almost to the gate of the town with the light infantry, rangers, Highlanders, and the grenadiers of the Royals, 15th, 17th, and 22nd Regiments.†

The "London Gazette," dated August 15th, 1758, says the troops actually with General Wolfe consisted of the grenadiers of the four oldest regiments, followed by the light infantry (a corps of 550, chosen as marksmen from the different regiments who served as irregulars), the companies of rangers, supported by the Highland regiment, and the remaining companies of grenadiers.

The garrison of Louisburg consisted of 3,000 regulars and burgher militia, and 350 Canadians and Indians.

The ordinary operations of the siege continued until the night of the 9th July, when the French (with five picquets, supported by 600 men) made a vigorous sortie, by which the grenadier company of the 17th Regiment was overpowered, its captain, the Earl of Dundonald, killed, Lieutenant Francis

* Capt. O'Callaghan's Account.
† "London Gazette," Aug. 15th, 1758.

Tew wounded and taken prisoner, and several men killed and wounded. The grenadiers of the 22nd and 28th coming up soon put an end to the fray. The French were at length driven back to the town, leaving 20 killed and about 80 wounded or prisoners. The enemy sent out a flag of truce to bury their dead, and when that was over the cannonading began again.*

The besiegers' works advanced steadily; a heavy fire was constantly kept up on the fortifications and shipping, and on the barracks and citadel, and some of the ships were burnt.

On July 26th, the governor surrendered to the British the fortifications and town of Louisburg, and the island of Cape Breton. Over 5,000 prisoners were taken, also 240 guns, 11 pairs of colours, and a large quantity of small-arms ammunition and stores, besides six ships of the line and five frigates burnt or destroyed in the harbour.

Between the 8th June and the 26th July, the regiment had the following casualties: Killed, Captain the Earl of Dundonald, 1 corporal and 10 privates; wounded: Captain Paul Ricaut, Lieutenant Francis Tew, 1 sergeant, 1 drummer and 31 privates.†

The British casualties on the whole were not severe, a little more than 500 of all ranks being killed and wounded during the entire siege.

The capture of Louisburg resulted in the transfer to British suzerainty of the islands of Cape Breton and Prince Edward's Island, and really heralded the conquest of Canada, as it deprived the French of the harbours, dockyards and arsenals, which formed the base of their power in North America.

During the period the regiment was at Cape Breton, a body of troops under Major-General Abercromby was repulsed at Fort Ticonderoga, on the west shore of Lake Champlain.

On the 21st July, a field order was published allotting awards to soldiers at the following rates, for such shot and shell as they might pick up, which had been fired by the enemy, viz.: For every 13-inch shell, a dollar; a 10-inch shell, half a dollar; an 8-inch shell, a quarter dollar. The shells to be brought to the mortar battery on the right. Large shot to be paid at 2d. each and smaller at 1d.; all arms taken

* "London Gazette," Aug. 15th, 1758.
† W.O 1, Book 1.

from the enemy to be brought to headquarters, 5s. being allowed for every good or repairable firelock.

North America, 1758.

On the 30th August, the 17th, brigaded with the 1st Battalion Royals, 48th, and Fraser's Highlanders, embarked at Louisburg, and sailing to Boston, marched through the woods to Lake George, where they joined the troops under Major-General Abercromby.

For the taking of Louisburg, July 27th, 1758, a medal, 1.7 inches in diameter, was struck in gold and silver and presented to a few of the officers who distinguished themselves.

Obverse: The British flag waving over the globe, marked "Canada," "America." On one side, a sailor waves his hat; on the other, a grenadier points to a female figure writhing beneath the globe, pointing to English boats approaching, and dropping the French lily into the sea; above, a flag, and a band inscribed "Pariter in bella." (Equal in wars.) Above hovers Fame with her trumpet and a laurel wreath. *Reverse:* Batteries firing; the English fleet in the offing; a French ship in flames and another towed away by English boats. Legend, Louisburg taken. MDCCLVIII.*† (See Plate 5.)

Another detail on the *reverse* shows the curved track of a bomb from an invisible mortar to the left of the battery, with the shell itself in the air about to alight on the town of Louisburg. The same medal in copper was given to the men.

The 17th went into winter quarters at Philadelphia.

1759

Canada, 1759.

In a return of troops serving in North America, dated, January 24th, the regiment is shown as having 660 rank and file, and 70 wanting to complete.

The month of May was taken up in preparation for the campaign by the lakes. Major-General Amherst arrived at Albany on the 3rd, and the greater part of the 17th arrived there on the 6th, when parties of the regiment and of the 42nd were ordered to take post at the rifts near Stillwater.‡ On the 22nd, the 17th were ordered up the River Hudson, and on the 25th, Colonel Darby, commanding the regiment, reported that Lieut. Watts, 17th, marching with a party of 12 men from

* "Medallic Illustrations," Vol. II., p. 685, No. 404.

† Photograph by the courtesy of Mr. A. D. Baldwin, Duncannon St., London.

‡ Major T. Mante, pp. 204, 206, 210; and General Gage's despatch.

Stillwater to the camp at Scorticook, had been attacked by 30 Indians when he (Lieut. Watts), and two men were killed, three wounded, and a corporal missing.

A later report by Major Hamilton stated his belief that the party was marching rather carelessly.*

A spirit of desertion now broke out among the troops, for which the General did not hesitate, for the sake of example, to approve all sentences of death, when such were awarded by general court-martial.

On the 3rd June, the rear of the army were ordered to take the field.

The drinking of spirituous liquors was totally discouraged, and a wholesome beverage was introduced in their stead, consisting of molasses and the tops of the spruce fir boiled together in a proper quantity of water, a mixture which had been found a most excellent antiscorbutic, and even an antidote against those distempers frequently occasioned by the excessive drinking of spirituous liquors.†

On the 8th June, the following field order was accordingly issued: "Spruce beer will soon be brewed for the army; it is hoped sufficient for the whole, and will cost the men but a very moderate price."‡

On the 6th June, the General had arrived at Fort Edward (on the east bank of the Hudson River, 56 miles from Albany), where he ordered the regular troops to assemble. Arriving on the 12th, they encamped in the following order: 1st Brigade; 1st, 55th, 27th; 2nd Brigade; 42nd, 77th, 17th; the provincials on the left of the 17th, the grenadiers and light companies of these regiments being formed in two battalions and encamped separately.

The troops then marched to Lake George, where a fort was erected, and boats procured to convey them along the lake.

On the 21st July, the boats having arrived, the General ordered the army to embark. This army consisted of the following:—

Royals, 17th, 27th, 42nd, 55th, 77th, amounting, officers included, to 5,279: total, 11,133.§ These were divided into

* C.O. 5, Book 54.

† Major T. Mante, pp. 204, 206, 210; and General Gage's despatch.

‡ J. Knox, Vol. I., pp. 360, 366, 367.

§ Major T. Mante, pp. 204, 206, 210; and General Gage's despatch.

four columns to cross the lake in the order laid down, the 17th being in the second column, commanded by Brigadier-General Gage, with 26 boats apportioned to it.*

Using blankets for sails, the troops arrived at the Second Narrows on the following morning.

Advancing towards Ticonderoga, they drove a body of French regulars and native Indians from a strong post two miles from the fort, and evinced such steady resolution, that the French commander quitted his fortified lines and embarked for Crown Point, leaving a garrison at Ticonderoga. The siege of this place was commenced; and on the 25th of July, the garrison blew up the fort and sailed to Crown Point, which place the French commander also abandoned, and retired down the lake to Isle-aux-Noix.

In General Amherst's daily diary of events, he mentions:

A false alarm on our left took place on the night of the 24th July, when a company of light infantry commenced firing, which was continued by some others, contrary to the General's orders, which were, that there was to be no firing at night, but that the troops were to stand fast, receiving the enemy with the bayonet. The result of the alarm was that an officer of the 17th was killed (name not mentioned) and some men wounded.†

Between the 22nd July (date of landing), and the 26th inclusive (date of enemy abandoning the fort), the regiment had two rank and file killed and eight wounded at the taking of Ticonderoga.‡

The British loss, in the reduction of the fort and lines, was one colonel, one lieutenant and 15 privates killed, and about 50 wounded.

On the 3rd August, news was received from Niagara, that the garrison there (numbering 607 and about 170 prisoners) had surrendered to Sir William Johnson on the 25th July.§

On the 4th, General Amherst arrived with the bulk of the army, to take possession of Crown Point, where a fort was erected, and a small naval force prepared for navigating the lake.

* J. Knox, Vol. I., pp. 360, 366, 367.
† C.O. 5, Book 56.
‡ J. Knox, Vol. I., p. 403.
§ C.O. 5, Book 56.

In a despatch from Major-General Amherst, dated, Crown Point, August 5th, he wrote: "A party I had sent to Crown Point brought in a deserter from the 17th, in a French coat, one whom I had pardoned for desertion when I was at Fort George. I thought it so necessary to make an immediate example, that I had him hanged directly." *

On the 8th August, General Amherst despatched Captain Quintin Kennedy (17th), and Lieut. Hamilton (Royals), with four Indians, on a special mission to General Wolfe at Quebec, Captain Kennedy having offered to go through the country by a much nearer way to the River St. Lawrence. On the 11th September, however, news was received that these officers had been made prisoners by some of the St. Francis Indians, who were hunting, and the officers were not released until the 15th November, when two privates of the regiment were handed over at the same time, who had been taken at Ticonderoga and Scorticook respectively.†

On the 11th October, the troops embarked and sailed down the lake in four divisions, with a view to attacking the enemy on the Isle-aux-Noix, but encountering high winds, and a frost having set in, they returned, reaching Crown Point on the 21st, and went into winter quarters.

General Amherst then proceeded to New York, where he arrived on the 11th December.

On the 13th September, when in command of the British forces at Quebec, there fell gloriously, in the hour of victory, Major-General James Wolfe, in whose brigade the 17th Regiment served at the siege of Louisburg.

His death has been thus described:—

"At the English right, though the attacking column was "broken to pieces, a fire was still kept up, chiefly, it seems, "by sharpshooters from the bushes and cornfields, where they "had lain for an hour or more. Here Wolfe himself led the "charge, at the head of the Louisburg grenadiers. A shot "shattered his wrist. He wrapped his handkerchief about it "and kept on. Another shot struck him, and he still advanced, "when a third lodged in his breast. He staggered, and sat "on the ground. Lieutenant Brown, of the grenadiers; one "Henderson, a volunteer in the same company; and a private

* "London Gazette." † C.O. 5, Book 56.

" soldier, aided by an officer of artillery who ran to join them,
" carried him in their arms to the rear. He begged them to
" lay him down. They did so, and asked if he would have
" a surgeon. 'There's no need,' he answered; 'it's all over
" 'with me.' A moment after, one of them cried out, 'They
" 'run; see how they run!' 'Who run?' Wolfe demanded,
" like a man roused from sleep. 'The enemy, sir. Egad,
" 'they give way everywhere!' 'Go, one of you, to Colonel
" 'Burton,' returned the dying man; 'tell him to march
" 'Webb's regiment down to Charles River, to cut off their
" 'retreat from the bridge.' Then, turning on his side, he
" murmured, 'Now God be praised, I will die in peace!'
" and in a few moments his gallant soul had fled."*

From the above account, it is clear that the grenadier company of the 17th was present with General Wolfe at Quebec on this occasion, the "Louisburg grenadiers" having consisted of the grenadier companies of the four oldest regiments with the force, viz., the Royals, 15th, 17th, and 22nd Regiments.† Lieutenant Brown referred to above was Lieutenant Henry Brown of the 22nd Regiment.

A Return of garrisons and winter quarters in North America, under Lieut.-General Amherst, dated, Headquarters, New York, 15th December, 1759, shows the headquarters and four companies of the 17th at Fort George, and six companies at Ticonderoga for the winter, the former place being considered of the greatest importance for keeping up an uninterrupted and safe communication with Ticonderoga.

Brigadier-General Forbes had died on the 11th March this year, and on the 24th October, the colonelcy of the regiment was conferred on Brigadier-General the Honourable Robert Monckton, from colonel-commandant of the 2nd Battalion 60th Regiment, one of the brave officers who chiefly contributed to the victory on the Heights of Abraham, on the 13th September, 1759, which, though it cost the life of the great Wolfe, gave Great Britain the possession of Quebec and Canada. Brigadier Monckton was wounded by a ball through his lungs, in making a charge at the head of Lascelle's Regiment, at the moment when the opposing generals, Wolfe and Montcalm, fell, and when victory declared for the English.

* From "Montcalm and Wolfe," by Francis Parkman; first published in 1884.
† Beckles Willson's "Life and Letters of James Wolfe" mentions the 45th also.

1760

The French possessions in Canada were invaded by the British troops in 1760 at three different points, the whole advancing upon Montreal: the First Division from Lake Ontario, the second from Lake Champlain, and the third from Quebec, up the River St. Lawrence.

The regiment formed part of the Second Division, under Colonel Haviland, which consisted of the grenadier and light companies of the 17th and 27th Regiments, under Colonel Darby, (17th); four companies, Royals, under Captain Mirrie; 17th Regiment, Major Campbell; 27th Regiment, Major Gordon; seven regiments provincials, five companies rangers, and a body of Indians. This division embarked at Crown Point* on the 11th of August, and, sailing towards Isle-aux-Noix, landed on the left bank of the River Richelieu, and captured a fort near the river: two other forts were abandoned by the enemy, and the British took possession of Isle-aux-Noix.

By September 5th, Brigadier Haviland had reduced Fort Chambli, where he found some of our brass field guns; and, on the same date, his advanced guard had arrived at Longueil, opposite Montreal. By a singular coincidence, the three armies reached the neighbourhood of Montreal on the same day.

The feeble and disheartened garrison now offered no resistance, and, on the 8th September, the French governor, being unable to withstand the forces opposed to him, surrendered Canada with all its dependencies to the British Crown.†

A Return of troops in North America, dated Quebec, 4th October, 1760, shows the 17th distributed between Schenectady, Fort Hunter, Fort Hendrick, Fort Schuyler, Fort Stanwix, and the east end of Oneida Lake.‡

On the 23rd December, two companies of the regiment, with two companies of the 22nd, were despatched to South Carolina, and took part in an expedition from Fort Prince George, under Lieut.-Colonel Grant (40th Regiment), against the Cherokee Indians. Owing to the extremely rough nature of the country, and the many difficulties to be encountered, it took some time to get in touch with the tribe.§

* J. Knox, Vol. II., p. 392. † Parkman, p. 126.
‡ W.O. 1, Book 5. P.R.O., London. C.O. 5, Book 56. § C.O. 5, Book 61.

1761

Brigadier-General the Honourable Monckton was this year appointed Governor and Commander-in-Chief of the province of New York, and promoted Major-General on the 20th February.

In an action with the Cherokees on the 10th June, the 22nd had an ensign killed, and the total casualties of the four companies (17th and 22nd) are shown together as eight privates.

In Colonel Grant's despatch, the four companies were commended for their "steadiness, regularity and coolness."*

N. America, 1761

In the summer, the regiment left its detached posts, and proceeded to New York, and in August, encamped at Staten Island.

When peace was settled with the Cherokees, the two companies of the 17th, mustering a total of 148, were ordered, in November, to Dominica, to join the expedition from New York, under General Monckton.*

The subduing of the French West India Islands having been decided on, an armament was assembled at Barbados, for the attack on them, and the land forces were placed under the command of Major-General the Honourable Robert Monckton, Colonel of the 17th Regiment. Lord Rollo had received orders to collect such troops as could be spared from the different garrisons in the West Indies, and with troops already under his command, to join the army under General Monckton, at Barbados.

The 17th (8 companies), sailed from New York on the 19th November, in the ships "Neptune," "Blackett," and "Duchess of Hamilton," mustering 488 of all ranks, and (included in the army from North America, under General Monckton) arrived on the 24th December at Carlisle Bay, Barbados, as did soon after Lord Rollo, with the corps under his command.

The armament consisted of 13,000 land forces, and a fleet of 16 sail of the line, besides many frigates, sloops of war, hospital ships, and transports ; and, sailing from Carlisle Bay, on the 5th January, 1762, proceeded against the island of

* C.O. 5, Book 61.

Martinique, which had been colonised by the French about the year 1635. After menacing the coast at several points, a landing was effected in the middle of January, in Cas des Navières Bay, when General Monckton formed his army on the heights above.*

Many difficulties were encountered from the rugged surface of the country, which was of itself a sort of natural fortification, and it was found necessary to commence operations by erecting batteries, to facilitate our approaches towards Mount Tortenson and Mount Garnier.

The following is a summary of General Monckton's despatch from the "London Gazette" of the 22nd March, 1762 :—

1762

Martinique, 1762.

On the 24th January, at daybreak, the troops advanced to the attack. Brigadier Grant, with the grenadiers, supported by Lord Rollo's brigade, began the attack on the outposts, under a brisk fire from our batteries. Brigadier Rufane marched along the shore, on the right, to secure the redoubts along the coast, whilst 1,000 seamen, in flat-bottomed boats, rowed up as he advanced. The light infantry under Colonel Scott, supported by Brigadier Walsh's brigade, marched on the left, to gain a plantation, with a view to turn the enemy's flank. Our designs succeeded in every quarter, and the enemy's works were now successively carried with an irresistible impetuosity, so that by 9 o'clock we were in full possession of Mount Tortenson.

Some of the enemy fled, in the greatest confusion, into the town of Port Royal, whilst others seized on Mount Garnier, which, being much higher than Mount Tortenson, overlooked and commanded it. Brigadier Haviland, with his brigade (two battalions of Highlanders and a corps of light infantry), had orders to cross the valley, to the left of Brigadier Walsh, to attack a body of the enemy on the opposite heights, and endeavour to get on their left, with a view to divide their force, but such was the difficulty of progress, that it was late before they effected their object, although they began their march at 2 a.m. On the 27th, about 4 p.m., the enemy

* Major T. Mante, pp. 351, 357.

descended from Mount Garnier and attacked the posts occupied by the light infantry and Brigadier Haviland's brigade, but they were received with the greatest steadiness and immediately repulsed. Our troops, here, were now reinforced by Brigadier Walsh's brigade and Brigadier Grant's division of grenadiers. Night had now come on, and Major Leland, advancing with his light infantry and finding no opposition, came upon, and took possession of, the enemy's redoubt, which (with the exception of a few prisoners taken) had been abandoned, the enemy's troops having fled into the town and citadel, and their militia dispersed in the country. Our light infantry were immediately supported by Brigadiers Walsh, Grant, and Haviland, so that by 9 p.m. on the night of the 27th we were in complete possession of Mount Garnier, and so precipitate was the enemy's flight, that they left a mortar loaded, and eight or nine guns, unspiked, with a quantity of ammunition and provisions. The guns and mortars were turned against the citadel in the morning. The General then ordered possession to be taken of Mount Capuchin, not more than 400 yards distant from the fort, resolving to erect batteries at both these places, in order to sooner reduce the citadel. The enemy, seeing extensive preparations, now judged it prudent to surrender, and beat the chamade on the evening of the 3rd February. On the following morning, the gate of the citadel of Fort Royal was delivered to His Majesty's forces, and, on the 5th, its garrison marched out to the terms of capitulation. It consisted of about 800 men, grenadiers, marines, militia, and freebooters, about 150 having been killed or wounded during the siege. In the several attacks, the enemy had not less than 1,000 men killed, wounded, and prisoners, among them some of the principal inhabitants. On the 7th, Pigeon Island was summoned to surrender, with which its commandant immediately complied.

The General now gave the necessary orders to repair the citadel, and was making the utmost despatch for attacking St. Pierre (the principal port of Martinique), when, on the 12th February, two deputies arrived, offering terms of capitulation of the whole island, on the part of the Governor-General.

Major-General the Honourable R. Monckton commended the conduct of the troops in his despatch, and added :—

" The difficulties they had to encounter, in the attack of " an enemy, possessed of every advantage of art and nature, " were great, and their perseverance in surmounting these " obstacles furnishes a noble example of British spirit."

In the Return of killed, wounded, and missing, at the attack on the enemy's redoubts, entrenchments, and breastworks between Cas des Navières, and Mount Tortenson, on 24th January, 1762, is shown :—Wounded : 17th light company, Captain Quinton Kennedy ; three rank and file killed, 16 rank and file wounded. In the attack on the villages and encampment before Fort Royal, one rank and file killed. (" London Gazette," 23rd March, 1762.)

General Monckton, as Commander-in-Chief of the Forces at Martinique, appointed Lieut.-Colonel John Darby, of the 17th, as Lieut.-Governor of it and its dependencies, and also as Governor of the fortress, called Fort Royal.*

The regiment had served throughout the siege in the 4th Brigade, and after the fall of Martinique our 5th Brigade, under Brigadier-General Walsh, proceeded to Grenada, to summon its governor to surrender, who, finding himself abandoned by the inhabitants, complied, without firing a shot.

On the 24th February, a threatened naval attack, by a small squadron, caused the surrender of the Island of St. Lucia, and, at the peace of 1763, the French agreed that the right to the island of St. Vincent should be vested in the English.†

THE FALL OF HAVANNAH.

Havannah, 1762.

As soon as it was definitely known, in 1762, that a rupture with the Court of Madrid was unavoidable, the British Ministry decided to avail themselves of the large body of land forces, then in the West Indies, by attacking the Spaniards, as they had the French, in some of their principal settlements; resolving to begin with Havannah, in the island of Cuba, which was looked upon as the key to the Spanish settlements in the West Indies.

* W.O. 1, Book 19; and " London Gazette," Aug. 25th, 1762.

† [It has been considered necessary to refer here to these islands, in order to correct a statement in Cannon's Record (at page xxviii. " Contents "), where he implies that sieges of these places actually took place, and that the regiment was present at them.]

An armament for the purpose had been placed under the command of General the Earl of Albemarle, the total of his force (exclusive of reinforcements from North America, and a detachment from Jamaica, which, together amounted to 2,000), being 12,041. General Monckton had the offer of the government of Martinique, or to go third in command in the expedition against Havannah, but his health having been greatly impaired from the West India climate, he chose to return to New York, to the government of which he had been appointed.

The Earl of Albemarle divided the army into five brigades, the 17th, under Lieut.-Colonel Campbell, being in the 4th Brigade, commanded by Brigadier-General Grant, and mustering 10 companies, with a total of 535 rank and file; the other corps in the brigade being the 1st and 2nd Battalions 42nd, and 77th Regiments.

This brigade accompanied the expedition through the Bahama Channel, and arrived within six leagues of Havannah on the 6th June; a landing was effected on the following day, and, on the 9th, the troops took up a position between Coximar and the Moro Fort, a strong fort of triangular form, having two bastions towards the land, and two irregular ones towards the sea, whilst it joined by a wall some strong batteries of heavy guns, which defended the entrance and overlooked the town.

The regiment took part in the services connected with the siege and capture of Moro Fort, which was the key position of the extensive works which covered the town.

The difficulties encountered in carrying on operations were particularly great; roads of communication had to be cut through thick woods, and the artillery had to be dragged a long way over a rough rocky shore. Several men on these services fell dead from heat, thirst and fatigue.*

Moro Fort was captured by storm on the 30th July, and, on the 11th August, a series of batteries opened so well-directed a fire on the defences of the town, that the guns of the garrison were soon silenced, and flags of truce were hung out.

The hardships which the English troops sustained, in forwarding their approaches to it, are altogether inexpressible. Though a great part of the provisions brought from England

* Major T. Mante, pp. 425-6, 461.

had been spoilt by the heat of the climate, the most distressing circumstance of the campaign was the scarcity of water. There being no river or spring near the troops, water had to be brought from a great distance, and, so scanty and precarious was the supply, that they were obliged to have recourse to water from the ships. On this occasion, excessive thirst soon caused the tongue to swell, extend itself outside the lips, and become black as in a state of mortification ; then the whole frame became a prey to the most excruciating agonies, till death at length intervened, and gave the unhappy sufferer relief. In this way, hundreds resigned themselves to eternity. A great number fell victims to a putrid fever. From the appearance of perfect health, three or four hours robbed them of existence.*

In the assault on the Moro, the English lost two lieutenants, and 12 rank and file killed, and one lieutenant, four sergeants, and 23 rank and file wounded. The loss of the Spaniards was 130 killed, 37 wounded, 310 prisoners, and 16 officers, all on shore, besides 213 drowned or killed in the boats—total, 706 men.

From our first landing in Cuba to the 13th August, this important conquest cost the English, in killed, wounded and prisoners, including those who died, 2,764 men.

The casualties in the 17th were Lieutenant Martin and Ensign McGrath, and two rank and file wounded ; one sergeant and two rank and file killed, and three rank and file missing ; four sergeants and 22 rank and file had also died.

In the Earl of Albemarle's despatch (" London Gazette," dated September 30th, 1762) he says : " The difficulties the " officers and soldiers have met with, and the fatigues they " have so cheerfully and resolutely gone through, since the " army first landed on the island, are not to be described ; " they deserve from me the highest commendation."

The capitulation was signed two days afterwards, and the British troops took possession of this valuable settlement, with nine ships of war in the harbour and two upon the stocks ; three ships of war were also found sunk at the entrance to the harbour.

* Major T. Mante, pp. 425-6, 461.

The preliminaries for a peace between France, Spain, and Great Britain, were signed at Fontainebleau on the 3rd November, 1762, which was proclaimed in the following year, Havannah being restored to Spain in exchange for Florida.

N. America, Aug., 1762.

The regiment left Havannah on the 20th August in three transports, named the "Hercules," "Duchess of Hamilton," and "Brotherly Love," and arrived at New York on the 24th.

The following medical report, on the state of health of the brigade (comprising the 17th, 1st and 2nd Battalions 42nd, and 77th Regiments) on return from Havannah, was rendered to General Amherst, by Surgeon John Adair, dated New York, 18th September, 1762:—

"I have visited the above regiments, and am sorry to "inform you of the deplorable situation they are in, being "reduced to the lowest state, with dangerous fevers and "fluxes, many of the men being past recovery and the rest "so weak, that I fear a long time will elapse before they are "again fit for service, their blood being in such a dissolved "and putrid state. Of the few I found capable of assisting "in the hospital as orderly men, the least fatigue has made "relapse, so that, of the whole brigade, there are not, in my "opinion, thirty men fit for service, nor are the officers in a "better condition; several cannot recover, and the greater "number of the remainder will, for a long time, be weakly, "and unfit to undergo much fatigue."*

A Recruiting Return, dated New York, for the year 1762, shows the recruits raised, from several American provinces, for our regular army, amongst which the 17th received 136 from the district of Virginia, 48 from North Carolina, and seven from Rhode Island.

A Return, dated 23rd December, shows Major-General the Honourable Monckton as Lieut.-Governor of the garrison of Annapolis Royal.

1763

Back Settlements of Canada, 1763.

Owing to a threatened rising, this year, amongst the Indians in the back settlements, especially at Detroit, it was considered necessary to send troops to frustrate their plans, which were, not only for the destruction of our most insignificant and remote forts, but also of our most important and central

* C.O. 5, Book 62.

settlements. General Amherst, therefore, confined his whole attention, at first, to Fort Pitt, Niagara, and Detroit.*

Light companies to our infantry regiments were not officially recognised and formed until early in 1771, but, throughout this campaign, General Amherst had found it advantageous to initiate (in addition to the existing grenadier companies), what were termed "light infantry companies" to line regiments.

Troops had already been despatched to various frontier posts, to quell the aggressive Indians, and letters from General Amherst, dated New York, 12th and 16th June, mention his sending the 17th Regiment, "such as it is (the West Indies "having greatly thinned it) to Albany, with orders to proceed "to Fort Stanwix, the light infantry company to embark at "New York on the 12th June, and it would be followed in a "few days by the rest of the regiment."

In a letter dated 2nd July, (with reference to dealing effectively with the mutinous Indians), General Amherst considered "our first attention was to make ourselves masters "of the lakes," a decision which was acted on by a campaign on the lakes in the following spring.

In a Return of the "Disposition of the Forces" in North America, dated 11th August, the regiment is shown as distributed between Oswego, west and east end of Oneida Lake, Fort Stanwix, Fort Edward, and Fort George.

According to Sir J. Amherst's intended distribution of the troops on Lake Erie, dated New York, 17th September, the detachments of the 17th and three other regiments were (on arrival of the 42nd at Detroit), to join their respective corps, if circumstances permitted, and, after the disbandment and drafting to another corps of one of the Independent Companies, its supernumeraries were to be sent to Oswego to join the 17th.

General Sir Jeffrey Amherst gave over command of the troops to Major-General Gage in November, this year, and returned to England. The distribution of the regiment in winter quarters was: Four companies at Fort Ontario and Oswego Falls, two companies at Fort Stanwix, one at Fort Edward, one at Fort George, and one divided between the posts at the east and west end of Oneida Lake.†

* Major T. Mante, pp. 481, 485, 507, 508.
† C.O. 5, Book 63.

1764

A despatch from General Gage, dated New York, 12th May, states :—The army now assembling for the service of the lakes, under command of Colonel Bradstreet (40th Regiment), will be composed of the provincial troops of New York (about 500), of Connecticut and Jersey together, nearly the same number, 300 Canadians, H.M.'s 17th, completed by drafts from the 55th Regiment, and four companies of the 80th Regiment, a detachment of 50 artillery with 10 guns—in all about 2,000.*

The plan of operations was, to send one body of troops to attack the different tribes of Indians settled between the Ohio and the lakes, under Colonel Bouquet, whilst another, under Colonel Bradstreet, was to attack the tribes on the banks of the lakes.

The latter force was ordered, in the large boats, and the Indians accompanying it, in their boats and canoes (all under the command of Lieut.-Colonel Campbell, 17th), to Niagara, where they arrived early in June, when Colonel Bradstreet, on arrival, assumed command of the whole.† Colonel Bradstreet's operations were to be made by water.

After leaving a sufficient garrison at Niagara, and, having been reinforced by a battalion of Canadians, and part of the 46th Regiment, he embarked at Lake Erie, on the 8th August, accompanied by 250 Indians, of whom nearly 100 belonged to Canada, and proceeded to Detroit, which he reached on the 26th August, having received, en route, deputations of submission from several tribes of Indians.

Thinking this a good opportunity, to take possession of the country of the Illinois (which had been ceded to the British by the peace of 1762), he ordered Captain Morris (17th) with proper instructions on that service, and, with him, an Indian of each of the different nations that accompanied his own party, and also a Frenchman as interpreter.

Colonel Bradstreet, on arrival at Detroit, ordered his troops to disembark. Its garrison, having sustained a long and severe blockade, was intensely gratified at its relief. The inhabitants were formed by him into three companies of militia, and the garrison was relieved by seven companies

* C.O. 5, Book 83.

† Major T. Mante, pp. 481, 485, 507, 508.

of the 17th, commanded by Lieut.-Colonel Campbell; two other companies of the regiment, with two companies of the new militia and a detachment of artillery, were ordered, under Captain Howard (17th), to take possession of Michillimackinac, which had remained unoccupied since its capture in the previous summer. Captain Howard effected his object without resistance, and, at the same time, sent parties of troops to re-occupy the deserted posts of Green Bay and Sault St. Marie. Thus, after the interval of more than a year, the flag of England was again displayed among the solitudes of the northern wilderness.*

Indians of various nations now began to demand audiences of Colonel Bradstreet. Letters received from Captain Morris showed that it was with great difficulty he had reached the fort on the Miamis river, where he met a number of savages, from whom he narrowly escaped with his life, as the account of the peace had not reached them. So much was he hampered by them, in fact, that he was forced to abandon his mission to the Illinois country, and returned to Detroit on the 17th September.

On the 14th September, Colonel Bradstreet left Detroit garrisoned by seven companies of the 17th and a detachment of artillery, the whole commanded by Lieut.-Colonel Campbell.†

1765

By advices from Michillimackinac, on the 6th January, Captain Howard (17th) had repaired the fort there, having completed two bastions and mounted two guns on them before the winter set in, and had procured a quantity of timber to provide the rest.

With reference to the submission of an influential Indian chief, a despatch from Lieut.-General Gage, dated, January 23rd, says:—

"There is nothing immediately necessary to be done "but to gain Pontiac, who has not been treated with, and "this treacherous savage, retaining his influence, may still "do mischief. On this account, orders have been sent to "Lieut.-Colonel Campbell, 17th Regiment, now commanding

* Parkman, p. 469.
† Major T. Mante, pp. 513, 516, 533.

" at Detroit, to send Pontiac a particular message, to come
" to him, and make peace for himself."

In a despatch from Lieut.-General Gage, dated, New York, 27th April, he states :—

" The Pottawattamies of St. Joseph's village recommenced
" hostilities very soon after their conference and treaty last
" year, with Colonel Bradstreet, at Detroit, by killing and
" scalping two men of the 17th, in November. The chiefs of
" the village had since been to Lieut.-Colonel Campbell, and
" were to meet him in the spring, and give the satisfaction he
" should require."

Another despatch, from Lieut.-General Gage, dated, New York, June 1st, relative to Detroit, reports that hostilities there had not ceased, a soldier of the 17th having been carried off by the Miami Indians on the 8th March.

A Return of Troops in North America, dated, New York, 24th September, shows seven companies of the regiment still at Detroit, and two at Michillimackinac, where they remained throughout the winter.*

1766

A Return, dated New York, 29th March, shows three companies of the regiment at Detroit, three at Niagara, two at Michillimackinac, and one at Oswego.

N. America, 1766.

In July, the headquarters of the regiment at Detroit, with detachments at out-stations, were relieved by the 2nd Battalion 60th, and left for New York, when the regiment was distributed as follows :—headquarters and five companies at New York, one company at Crown Point, one at Ticonderoga, one at Albany, and one divided between Fort Stanwix and Fort George.

A letter from Lieut.-General Gage, dated, November 11th, states, that about a fortnight previously, the company of the 17th at Albany was called out one night, to quell a rising on the part of a mob, who had pulled down a storehouse belonging to the British, and robbed it of a quantity of provisions.

The regiment remained distributed as above, until it was concentrated at New York, in July the following year, prior to embarking for England.†

* C.O. 5, Book 83.
† C.O. 5, Book 84.

CHAPTER V.

HOME SERVICE, NORTH AMERICA, AND NOVA SCOTIA.
1767–1785.

1767

To England, 1767.

ON the 24th July, the regiment (nine companies) embarked at New York for England, and sailed for Portsmouth. A War Office order directed it on disembarking to march as follows: Three companies to Bridgewater, two to Wells, two to Somerton, one to Shepton Mallet, and one to Taunton.

1768

A letter, dated 5th February, directed that the regiment, being so weak in numbers, was not to be reviewed (inspected) this spring.

In August, the whole regiment moved to Taunton, marching in September to Chatham, and in October to Maidstone, with detachments at Dover and Upnor Castles.

1769

The regiment having returned to Chatham, was reviewed there on the 17th May by Major-General Carey. Amongst the inspecting officer's remarks was: "The attention and care "of the officers are, in every particular, highly commendable, "and, considering the number of recruits, the regiment "performed remarkably well."

On the 23rd May, the regiment received orders to march from Chatham as follows: Headquarters and five companies to Kingston and Hampton Wick, and four to Tooting, Mitcham, Merton and Wimbledon, leaving detachments at Dover and Upnor Castles.

On the 8th June, the King, "attended by several general "officers, and the nobility," reviewed on Wimbledon Common the 2nd Battalion of the Marquis of Lorne's Regiment (1st Royals), and also Major-General the Honourable Monckton's (17th) Regiment.*

Between the 9th June and 16th September, the headquarters and five companies had been directed to march from Kingston to Dover Castle and back to Chatham, where the regiment was again concentrated by the 12th March in the following year.

1770

On the 13th March, the regiment, leaving a detachment at Dover Castle, was ordered to march: Six companies to Newcastle, detaching one company to Sunderland and one to Tynemouth.

The headquarters were reviewed at Newcastle, on the 11th June, by Major-General Murray; under arms, 308 rank and file.

Amongst the inspecting officer's remarks were:—

(1) "The regiment having so many young men in it, is "not fit for active service."

(2) "Nothing can exceed the zeal and attention of the "officers, and everything may be expected from a corps like "this, in which the officers vie with each other to promote the "good of the service."

Major-General the Honourable Monckton was this year promoted Lieutenant-General.

1771

On the 11th January, an augmentation of 180 privates was ordered, and a light company was also added to the regiment, consisting of three sergeants, three corporals, two drummers, and 62 private men.

The newly augmented establishment was to be: One lieut.-colonel, one major, eight captains, twelve lieutenants, eight ensigns, one chaplain, one adjutant, one quartermaster, one surgeon, one surgeon's mate, 20 sergeants, 30 corporals, ten drummers, two fifers to the grenadier company, and 380 private men (ten companies), making a total of 477, to take effect from the 26th March.

* "Lloyd's Evening Post," 8th June, 1769, and "St. James's Chronicle," 8th June, 1769.

The sum of five guineas, which had been allowed from the 25th December, 1770, to recruiting officers (half of this being given to the recruit on enlistment), was changed in March, 1771, to three and a half guineas for officers, and one and a half for recruits.

In the early part of this year, a considerable correspondence took place respecting the poundage on the soldiers' pay. This poundage—a shilling in the pound—was to be taken out of a fund given by the King's bounty to the agents of the regiments, and was to be subjected to drafts by the regimental paymaster as might be required. The fund was also to be used to reimburse the rank and file for the deductions made upon them for the paymaster and surgeon. The saving to the soldier by this fund was calculated to be about 16s. 6d. per annum.

Scotland, 1771.

On the 2nd April, the regiment was ordered to march from Newcastle to Berwick, and thence to Edinburgh, where it was inspected on the 5th June by the Duke of Argyll, and described in the report by the inspecting officer as "a remarkably "fine corps, well clothed, well dressed, and well disciplined."

1772

On the 18th August, a Royal Warrant amended the regulations as to the prices of officer's commissions.

The regiment was inspected this year, at Edinburgh, by Major-General Oughton, on the 11th May, and reported on as "a very fine corps, perfectly well trained, and fit for "immediate service."

1773

Ireland, 1773.

It left Scotland early in February, embarking on the 10th at Port Patrick for Ireland, where it landed the same day, at Donaghadee, and marched to Dublin. The Embarkation Return shows that the headquarters mustered a total of 338 of all ranks, and a footnote to it states: "The Baggage, Sick and Lame "embarked at Irvine, but the numbers had not been returned."

In a book of "Irish Martial Affairs,"* an order, dated 27th May, shows the regiment receiving the additional allowance, then granted, for doing duty in Dublin from the 23rd February to 26th May inclusive.

*Vol. XLI., p. 108.

The inspection this year took place in the Phœnix Park, Dublin, on the 18th May, by General M. O. Dilkes, and shortly after the regiment marched to Kilkenny. The Inspection Report says: "The officers made a very fine appearance, and "all the movements were performed with great attention."

1774

It was inspected at Kilkenny on the 28th May, by Major-General the Earl of Drogheda, and, in June, marched to Clare Castle.

The Inspection Return shows 376 effective rank and file, and the officers wearing "Silver Swords," apparently meant for silver sword-hilts.

1775

The Monthly Return for June, shows the headquarters at Galway, with a subaltern's detachment at Auchterard.

The inspection, this year, took place at Galway on the 9th June, by Major-General Gisborne, after which the regiment marched to Athlone. The inspecting officer's report shows it "in great order and very fit for service."

Serious disputes had now arisen between the British colonists in North America and the government, and the colonists evinced a daring spirit of resistance in their opposition to the measures for raising a revenue in their country, which, in April, 1775, was followed by open hostility, some provincial militia firing on a detachment of the King's troops, on its march from Boston to Concord, to take possession of a quantity of military stores at the latter place. This was followed by the assembling of multitudes of armed men near Boston; and when the news of these occurrences arrived in England, several regiments were ordered to embark for America. The 17th Regiment, which had moved to Athlone, was directed to hold itself in readiness for service abroad, and on the 4th August it embarked in Ireland for North America.

North America, 1775.

Its strength was: Ten companies, each consisting of two sergeants, three corporals, one drummer, and 38 privates—477 of all ranks.*

* W.O. 1, Book 610.

On Dec. 31st, the colonists made an assault on Quebec, when one of their leaders (General Montgomery) was killed.

Richard Montgomery,* born on December 2nd, 1736, in Ireland, came of an ancient French family. His father, a baronet, and once an M.P., gave him a good education in Dublin, and he was gazetted Ensign in the 17th Foot on the 21st September, 1756, becoming Lieutenant 10th July, 1758, and Captain 6th May, 1762, retiring from the British army in April, 1772, and, during his service, had fought under General Wolfe at Louisburg, under General Amherst at the capture of Ticonderoga and Crown Point, and, presumably, in many other actions in which the regiment was engaged up to the date of its return home in July, 1767.

On retirement, he emigrated to America, purchased a fine estate on the banks of the Hudson River, and, marrying the daughter of R. R. Livingstone (a man of note), settled in New York, became a famous American general, and served with the revolutionary army.

Towards the revolted colonists there was at this period great sympathy at home, which existed in the highest ranks of the navy and army, as can be seen by reference to Sir George Trevelyan's "History of the American Revolution." †

So great was the esteem in which General Montgomery was held by the British, that, on the occasion of his death in action, his body, by General Carleton's order, was brought into the town of Quebec, with every mark of reverence and regret, and buried with military honours, it having been borne by our soldiers to a new grave in the gorge of the St. Louis Gate.‡

From the high praise accorded to his merits in American histories, it is clear that, as a born leader of men, General Montgomery was worshipped by the Americans, and apropos of the inspiring effect his presence had on those under him, one writer expressively says: "Men would follow him without "knowing why."§

1776

The regiment was detained some time by contrary winds, but it landed at Boston on the 1st of January, 1776. At this

* Portrait of General Montgomery reproduced by the courtesy of Messrs. Stevens and Stiles, 39, Great Russell Street, London, W.

† Vol. I., pp. 85-6, Vol. II., pp. 193-4 and 225-6, Vol. III., p. 385.

‡ A brass tablet now marks the spot, near the present St. Louis Gate. "Old Quebec," G. Parker, p. 359.

§ "Our Struggle for the 14th Colony," p. 370, by Justin H. Smith

period the British troops at Boston were environed on the land side by a numerous army of provincials; much inconvenience was experienced in procuring provisions, and as this town did not appear to be a place calculated to become the base of extensive military operations for the reduction of the revolted provinces, Lieut.-General Sir William Howe resolved to vacate Boston, and proceed with the army to Nova Scotia; this resolution was carried into effect in the middle of March, when the 17th sailed with the army to Halifax.

Reinforcements being expected from England, the army sailed from Halifax in June, and, proceeding to the vicinity of New York, landed, on the 3rd of July, at Staten Island, where the 17th, commanded by Lieut.-Colonel Mawhood, was formed in the 4th Brigade with the 40th, 46th and 55th Regiments under Major-General James Grant.*

On the 22nd of August a landing was effected by Sir William Howe, without opposition, on Long Island, and on the evening of the 26th, the army was put in motion to pass a range of wooded heights, which intersect the island, and attack the American army in position beyond the hills. The 17th Regiment formed part of the column under Major-General Grant, which was directed to advance along the coast, with ten guns, to draw the enemy's attention to that quarter. Moving forward at the appointed hour, this column fell in with the advanced parties of the Americans about midnight, and, at daybreak on the following morning, encountered a large force, formed in an advantageous position defended by artillery. Skirmishing and cannonading ensued, and was continued until the Americans discovered, by the firing at Brooklyn, that the left of their army had been turned and forced, when they retreated in great confusion through a morass. They were met and attacked by the Second Battalion of grenadiers, which was soon reinforced by the 71st Highlanders, and were also assailed on the left by Major-General Grant's corps, and sustained severe loss, many of the Americans being killed, and others drowned or suffocated in the morass. The American army was driven from its positions with a severe loss of upwards of 2,000, and made a precipitate retreat to the fortified lines at Brooklyn.

* C.O. 5, Book 93.

The prisoners taken from them in the course of the day amounted to 997 men, of whom nine officers and 58 privates were wounded. They also lost five guns and one howitzer.*†

The regiment had Captain Sir Alexander Murray, and two rank and file killed; Lieutenant Marcus Morgan, one sergeant, and 19 rank and file wounded.

The Americans having quitted their fortified lines at Brooklyn, and passed the river to New York, the conquest of Long Island by the British troops was completed; and the 17th Regiment shared in the operations by which the capture of New York was accomplished; also in the movements by which the Americans were driven from White Plains; and in the reduction of Fort Washington. Afterwards proceeding to the Jerseys, the regiment was stationed at Brunswick, and subsequently at Princetown.

During the winter, General Washington suddenly passed the Delaware River, and surprised and made nearly a thousand prisoners of a corps of Hessians at Trenton, who, from an idea that the enemy were too much weakened to attempt anything serious, had sunk into a false security, productive both of want of discipline and attention. After this *coup-de-main*, he retired. Being reinforced, he again passed the river, and took up a position at Trenton, which was considered on both sides as a post of the first importance to secure possession of.‡ Major-General Earl Cornwallis advanced with a division of British troops, and, after reconnoitring the American position, sent orders for the 17th, 40th, and 55th Regiments to join him from Princetown. These regiments constituted the 4th (Grant's) Brigade, the command of which temporarily devolved on Lieut.-Colonel Mawhood (17th),§ and he also had with him two guns and three troops of light dragoons, in all over 700 men.

* Beatson's Memoirs.

† R. Lamb's "Journal of the American War," states that, in this action, the Americans lost upwards of 3,000 men, including nearly 1,100 prisoners and 32 guns. Among the Americans who fell, a regiment from Maryland was particularly regretted. It consisted wholly of young men of the best families in that province, all of whom behaved with the most admirable heroism, and were all killed or wounded.

‡ MSS. Records R.U.S.I., London.

§ W. S. Stryker, p. 248, and Rev. Dr. Belcher, p. 204.

1777

On the morning of the 3rd January, General Washington gave Lord Cornwallis the slip. In order to prevent any suspicion of his design, he ordered his camp fires to be kept up and his patrols to go about as usual, and marched with his army a little after midnight on the 2nd, in the hope of surprising a brigade of the King's forces encamped at Princetown. To avoid falling in with Lord Cornwallis's post at Maidenhead, which lay in the direct route to Princetown, he took a circuitous route, and, soon after daylight on the 3rd, met the very brigade, of the regiments above mentioned, he had intended to surprise.*

Early that morning the three regiments had commenced their march. The 17th (only 240 strong), commanded by Lieut.-Colonel Mawhood, being in advance, encountered the van of the American army. The morning being foggy, Lieut.-Colonel Mawhood could not discern the numbers of the force he had met; but supposing it to be only a detachment, he instantly attacked his opponents, and the 17th speedily drove back a force of very superior numbers with great gallantry. The regiment was soon surrounded in front and on both flanks by a numerous force; and Lieut.-Colonel Mawhood, discovering that he was engaged with the right wing of the American army, resolved to make a desperate effort to extricate himself: having confidence in the valour and resolution of the regiment, he directed a bayonet charge to break through the American army. Undismayed by the number of their opponents, the 17th rushed upon the ranks of the enemy, and driving them back to a ravine, which separated them from their rear,† continued their march to Maidenhead. Their conduct excited great admiration, and the Americans acknowledged the superior gallantry of the regiment, in saying that "nothing could exceed the gallant "behaviour of the 17th." Since the battalion only mustered just over 200, this feat was rightly estimated as one of the most gallant exploits of the war.‡

A serious loss was, however, sustained, 101 out of 240 officers and men being killed, wounded and missing. Among

* Beatson's Memoirs, Vol. IV., p. 195.
† Mr. Richards' MSS.
‡ MSS. Records, R.U.S.I., London.

the former was Captain the Honourable William Leslie, second son of the 6th Earl of Leven and Melville, an officer of great promise, whose death was much regretted (see Plate 9); also 12 rank and file killed; one captain, one lieutenant, one ensign,* four sergeants and 46 rank and file wounded, and one sergeant, one drummer, and 33 rank and file missing. Many of the latter joined later.

Captain the Honourable Leslie was mortally wounded in the fight, and when discovered by General Washington, as the latter passed over the field after the battle, was properly cared for by Dr. Benjamin Rush, who was with Washington that day. Dr. Rush attended to the wants of his wounded foe with more than ordinary interest, in return, as he told General Washington, for some obligation which he owed to Captain Leslie's father, for many kindnesses received at his hands when a student at the university in Edinburgh. Captain Leslie was carried off by the army on their march northward, and received every possible attention, but he died the next morning near Pluckemin, and on the following day, January 5th, was buried with military honours in the village cemetery at Pluckemin.

Dr. Rush further showed his regard for the father of the young officer by erecting a monument to Captain Leslie's memory in the graveyard at Pluckemin.

The following is the inscription thereon:—

In Memory of the
Honourable Captain William Leslie,
of the 17th British Regiment,
Son of the Earl of Leven,
in Scotland.
He fell January 3rd, 1777,
Aged 26 Years,
at the Battle of Princetown.
His friend, Benjamin Rush, M.D.,
of Philadelphia,
hath caused this Stone to be
erected as a Mark of his esteem
for his WORTH, and of his respect for
his noble family.†

The despatch of General Sir William Howe, dated New York, January 5th, 1777, states: "The bravery and conduct,

*The "London Gazette," dated 22nd Feb., 1777, unfortunately does not mention the names of the wounded officers, nor does any of the official correspondence.

† W. S. Stryker, p. 292.

" particularly of the 17th, are highly commended by Lord
" Cornwallis," and continues :—

" His lordship, finding that the enemy had made this
" movement, and having heard the fire occasioned by Colonel
" Mawhood's attack, returned immediately from Trenton;
" but the enemy being some hours' march in front, and keeping
" this advantage by an immediate departure from Princetown,
" retreated by King's Town, breaking down the bridge behind
" them, and crossed the Millstone River at a bridge under
" Rocky Hill, so as to occupy a strong country.

" Lord Cornwallis, seeing it could not answer any purpose
" to continue his pursuit, returned with his whole force to
" Brunswick, the troops upon the right being assembled at
" Elizabeth Town."

GENERAL HOWE'S CONGRATULATIONS.

" Headquarters, New York,
" January 8th, 1777.

" General Howe desires Lieut.-Colonel Mawhood will
" accept his thanks for his gallantry and brave conduct in the
" attack on the enemy on the 3rd instant. He desires his
" thanks also may be given to the officers and soldiers of the
" 17th Foot, to part of the 55th Regiment, and other detach-
" ments on the march, who, on that occasion, supported the
" 17th Regiment, and charged the enemy with bayonets in the
" most spirited manner. The General desires his public
" approbation may be signified to Captain William Scott of
" the 17th Foot, for his remarkable conduct in protecting and
" securing the baggage of the 4th Brigade on the above
" occasion."

The following message was received from the King, through Lord George Germain, principal Secretary of State for America, dated, Whitehall, 3rd March, 1777 :—

" His Majesty has been pleased to take very particular
" notice of the bravery of Lieut.-Colonel Mawhood, and
" approves the behaviour of the regiments under his command,
" especially the 17th, so highly commended by Lord Corn-
" wallis."*

* C.O. 5, Book 94.

"The bravery and abilities of Colonel Mawhood, on this "occasion, deservedly gained him the highest applause;"* and the resolute attack of the 17th so occupied the American army that the 40th and 55th Regiments effected their retreat with much less loss than could have been expected. The American army had many men killed and wounded on this occasion; among the killed was an officer of reputation, Brigadier-General Mercer from Virginia.

Referring to Colonel Mawhood's celebrated action, on the 3rd January, 1777, the author of "The American Revolution" (Sir George Trevelyan) gives the following account, at pages 145–8, Vol. II. :—

Page 145 :—

In the van of the American army was a weak brigade of continental infantry, under the command of General Mercer.

Page 146 :—

Both parties raced for the possession of an orchard, which lay midway between them, which the Americans reached first. Three volleys were exchanged at a distance of 40 yards, and then Colonel Mawhood led his regiment with a bayonet charge. The continental soldiers broke and fled, but some of the officers remained at their posts, and died very staunchly. Two New Jersey field-pieces were captured, and the captain in charge of them was killed at his guns. General Mercer himself used his sword until he fell, covered with wounds. Those who witnessed the behaviour of the 17th Foot on that occasion might well ask themselves, what would have happened if Lord Cornwallis had hurled, not one, but twelve or fifteen of such regiments against the right wing of the American army, while it was enclosed and entrapped between the ice-laden flood of the Delaware, and the unfordable Assumpink Creek.

The British followed in pursuit, but found themselves in presence of numerous reinforcements, which were flocking in, . . . and the whole space in front and flank of the English regiment was rapidly thronged with as many militiamen, regulars, and riflemen from the western frontier as could find room to ply their firelocks.

*Beatson's Naval and Military Memoirs.

Lieutenant G. Peevor.
1810.

Captain The Honourable William Leslie.
1777.

Lieut. W. H. J. Disbrowe.
1856.

The adversaries were separated by so short a distance that they could hear each other speak, during the moments which elapsed before the roar of musketry commenced.*

The line of British infantry, a bare 400 to begin with, must very soon have been annihilated. No military object could be promoted by such a tragedy, and enough having been done for honour, Colonel Mawhood turned his attention to saving the remnant of his battalion. He abandoned the two cannon he had taken, and two others of his own, and made off in the direction of Trenton, covering his retreat, as best he might, with a handful of cavalry.†

A "Distribution of Troops," dated, New York, 8th May, shows the 17th at New York, in the 4th Brigade, under Brigadier-General Agnew, with the 37th, 46th and 64th Regiments.‡

When the army took the field, the regiment was employed in operations in the Jerseys to bring the American army to a general engagement, but General Washington kept close in his strong position in the mountains; and the British undertook an expedition to Pennsylvania: the 17th were employed in this enterprise, and were formed in brigade as above.

A landing was effected on the northern shore of Elk River on the 25th of August, and the army of the revolted provinces took up a position at Brandywine Creek to oppose the advance; an attack was made on the position on the 11th of September, when the Americans were driven from their ground with loss. On this occasion the 17th formed part of the column under Major-General Earl Cornwallis, and had no casualties.

On the 25th September, the army marched in two columns to Germantown, about six miles from Philadelphia, and there encamped. Lord Cornwallis, with the British grenadiers, and

* As the 1st Virginians were being got into position, Captain John Fleming called out: "Gentlemen, dress the line." "We will dress you," a British private retorted, and Fleming was killed the next moment.

† "In this trying and dangerous situation, the brave commander, and his equally brave regiment, have gained immortal honour." That sentence, from "The History of Europe," in the "Annual Register," expressed the unanimous opinion of Colonel Mawhood's countrymen.

‡ C.O. 5, Book 94; and "London Gazette," 2nd December, 1777; and Mr. Richards' MSS.

two battalions of Hessian grenadiers, took possession of Philadelphia next morning.*

The following is recorded from a short biography of Captain Brereton, who was then serving in the regiment :—

The day after the capture of Philadelphia, two American frigates and sixteen gunboats came up the Delaware River to drive out the English.

One frigate, the "Delaware" (30 guns), got ashore, and was attacked by the 17th grenadiers, under Captain Brereton, and two guns of the battery. She struck, and Brereton, taking his artillery on board, turned the guns against the enemy. Then commenced a general fire from the guns and grenadiers, which, with the assistance that the captured frigate afforded, soon obliged the whole force of the enemy to retreat down the river in confusion.†

Lord Cornwallis, commanding at Philadelphia, immediately sent his thanks to Captain Brereton, commending him for his valour.

(The above incident is referred to in the "London Gazette," dated December 2nd, 1777, but Captain Brereton's name is not mentioned.)

He was again engaged, and succeeded in an exploit, which was a matter of much remark at the time, in having (at Province Island, near Philadelphia) re-taken a battery, erected against Mud Island, which had been previously given up as lost by a senior officer. This was achieving a most important step, for it is impossible to calculate how much mischief might have been done, had the enemy made themselves complete masters of the island. Captain Brereton again received the thanks of Lord Cornwallis, and was left in command, its former commanding officer being tried by court-martial, and dismissed the service.

The Americans attempted to surprise the British troops early on the 4th October, and gained some advantage at first, but were speedily repulsed with severe loss. On this occasion, six companies of the 40th Regiment, commanded by Lieut.-Colonel Musgrave, occupied a store house, which they most gallantly defended, when attacked by an American brigade,

* Beatson's Memoirs, Vol. IV., p. 259.

† C.O. 5, Book 94; and "London Gazette," 2nd December, 1777; and Mr. Richards' MSS.

until Major-General Grey, with the 3rd Brigade, and Brigadier-General Agnew, with the 4th Brigade covering his left, repulsed, by a vigorous attack, the enemy, who had penetrated into the upper part of the village, and this was done with great slaughter.

Earl Cornwallis, who had been apprised at Philadelphia of the enemy's approach, moved out with two British and one Hessian division of grenadiers, and a squadron of dragoons. Getting to Germantown just as the enemy had been forced out of the village, and joining Major-General Grey, he headed the troops and followed them eight miles, but they retreated so quickly he was unable to overtake them.

In this action, the enemy had upwards of 200 killed, about 600 wounded, and nearly 400 prisoners.

Since the battle of Brandywine, 72 of their officers had been taken, exclusive of 10 belonging to the "Delaware" frigate.

The King's army had about 100 killed and wounded at Germantown. The 17th had Ensign Nathaniel Philips, one sergeant, and four rank and file killed; three sergeants and 21 rank and file wounded.

In the different skirmishes from the 4th to 8th December, the regiment is shown as having three rank and file missing, and also Lieutenant Matthew Anketell wounded, whilst serving with the 1st Light Infantry.

The regiment passed the winter in quarters at Philadelphia.

1778

In a "State of the Forces," commanded by General Sir William Howe, dated Philadelphia, 24th March, the regiment is shown in the 3rd Brigade, with the 15th, 42nd, and 44th Regiments, the 17th mustering 17 officers, four staff, and 270 effective rank and file; also five sergeants, three drummers, and 57 rank and file, prisoners with the rebels.*

In the spring of this year the regiment furnished several detachments, which ranged the country in various directions around Philadelphia, to open communications for bringing in supplies, and to collect forage for the army. A despatch from Sir William Howe, dated Philadelphia, 11th May, states:

* C.O. 5, Book 95.

"These detachments have, without exception, succeeded to "my expectations; Colonel Mawhood (17th) in particular, "with three battalions made a descent, in March, on part of "Jersey near Islam, and after dispersing the force assembled "there, returned with a very seasonable supply of forage."

The regiment took part in the fatigues and difficulties of the march from Philadelphia through the Jerseys, in order to return to New York, and its flank companies were engaged in repulsing the attack of the rebel army, on the rear of the column, at Freehold, New Jersey, on the 28th June, on which occasion Captain William Brereton, commanding the grenadier company, was wounded, one private died from fatigue, and the other casualties were two privates missing.

The regiment, on its return to New York, in July, was quartered at Long Island, and a Return, dated 1st November, shows the above brigade broken up, and the regiment at New York, its number of prisoners with the rebels having been now reduced to one sergeant and 12 rank and file.*

On the 14th November the regiment was in winter quarters at Kingsbridge, New York Island.

Its establishment was augmented by one corporal and 14 privates per company, from the 25th December this year.†

1779

A Report from Major-General Wm. Tryon, dated Fore Post, Kingsbridge, 28th February, states: "In accordance "with instructions received, he had moved out of the lines "before 11 p.m. on the 26th instant, with the 17th, 44th and "57th Regiments, a Hessian regiment and their guns, and "pushed on to Horse Neck, whence, on our approach, the "rebels, after discharging a few rounds of grape shot, fled, "abandoning their guns and the town. At 10 a.m. on the "27th, the 17th and 100 men of the 44th were detached, "under Lieut.-Colonel Johnson (17th) to Greenwich, to "destroy the salt works and ships there, which was effectually "carried out by a large storehouse (containing fine salt and "26 cast-iron salt pans) having been destroyed, and a fine

* C.O. 5, Book 96.
† W.O. 1, Book 10.

" schooner and two sloops in the harbour having been burnt.
" At Horse Neck we destroyed a quantity of military stores
" and provisions and three of the rebel guns ; also a few rebels
" were killed, and 26 prisoners taken. This force had returned
" to Kingsbridge by 4 p.m. on the 28th, having been out
" forty-one hours, and undergone, with great perseverance and
" cheerfulness, the fatigue of a march between fifty and sixty
" miles, over uncommonly bad roads, and with partly un-
" favourable weather."*

A Monthly Return, dated 1st May, shows the regiment still at Kingsbridge, and, by the 15th June, it had moved to Stoney Point, a fortified post on the River Hudson, where it was placed in garrison with the grenadiers of the 71st, a company of loyal Americans and a detachment of artillery—in all 600, under command of Lieut.-Colonel Henry Johnson (17th).

At 1 a.m. on the 16th July this post was suddenly attacked by 1,200 American infantry, under General Wayne, whose force, being divided into two columns, entered the works in opposite directions, and met in the centre. The 17th made a gallant resistance, but were overpowered by superior numbers. General Wayne was slightly wounded in the head by a musket ball. The Americans carried the fort with the bayonet without firing a shot, with a loss of 98 killed and wounded.†

In the "London Gazette" of October 27th, 1779, a despatch from General Sir Henry Clinton states that, on hearing of it, he pushed forward a necessary force for the recovery of Stoney Point, under Brigadier-General Stirling, when, upon the latter arriving within sight of Stoney Point, the enemy made a precipitate retreat, and with some instances of disgrace, whereupon Brigadier-General Stirling occupied the fort and remained there for a while with five battalions, to repair the works, which were a good deal damaged.

Copy of a letter from Lieut.-Colonel Johnson, 17th Foot (late commanding troops at Fort Stoney Point) to General Sir Henry Clinton, dated Hardy's Town, July 24th, 1779 :—

" The bearer, Lieutenant Armstrong of the 17th Infantry,
" will give you a full and perfect account of the unfortunate
" event of the morning of the 16th instant, when the post of

* C.O. 5, Book 98.
† Avery, Vol. VI.

" Stoney Point fell into the hands of the enemy. I am inclined " to think that, upon a just representation, you will be fully " convinced that it was not any neglect on my part, nor of " the troops under my command, but the very superior force " of the enemy that caused the capture of the place.

" Enclosed I send a Return of killed, wounded, missing " and prisoners, as near as could be collected by commanding " officers of corps."

17th, killed : Captain Tew, two sergeants, and 16 rank and file. Wounded : Lieutenant Simpson, Ensign Sinclair, and 43 rank and file. Missing : One drummer and 20 rank and file.

The total number of prisoners taken was : 27 officers, one conductor, and 590 men,* amongst which the 17th had the following, viz. : Lieut.-Colonel Johnson, Captains Darby and Clayton, Lieutenants Armstrong, Carey, Williams, Simpson and Hayman, Ensigns Sinclair and Robinson, Adjutant Hamilton, Surgeon Horn, 17 sergeants, 12 drummers and 222 rank and file.†

Captain Tew's death was universally lamented. He had frequently been left for dead on the field, and could show more wounds, received in the German and American wars, than any other officer in the British, or perhaps any other service. Promotion had long been promised him, but it was his fate to finish his honourable career at the head of a company.

1780

Monthly Returns, up to the 15th August this year, show officers and men of the headquarter detachment still prisoners with the rebels, and in the interval between the siege of Stoney Point and the above date, a captain's detachment of the regiment, with two ensigns, the quartermaster, surgeon's mate, and less than 100 men fit for duty had been quartered at Long Island, New York.

This detachment was united with a number of others, of provincial troops, and placed under the orders of Colonel Watson of the Foot Guards, and they all embarked with the Brigade of Guards, a detachment 82nd Regiment, Bose's

* C.O. 5, Book 98.
† C.O. 5, Book 98.

(German) Regiment, and the King's American Regiment of Provincials, for service in Virginia, under Major-General the Honourable A. Leslie, and sailed from New York on the 16th October. General Leslie's instructions, on proceeding to Chesapeake Bay, were, to make a diversion in favour of Lieut.-General Earl Cornwallis, who by the time of General Leslie's arrival would probably be acting in the back parts of North Carolina.* During the siege of Gibraltar, which was now going on, the death of Colonel Mawhood, (who had only recently joined the 72nd Regiment there), made the 29th August a mournful day. He was respected for his good services.†

An official Return shows the following number of officers who were exchanged as prisoners of war since the 25th October, 1780, viz.: One major-general, six lieut.-colonels, seven majors, 29 captains, 57 lieutenants, 24 ensigns, one cornet, and seven 2nd lieutenants (R.A.), amongst which the head-quarter detachment of the 17th was included, and on release was quartered at Long Island.‡ In the meantime, the detachment with General Leslie's force was encamped at Portsmouth, Virginia, in front of the redoubts which covered the town.

1781

A despatch from Earl Cornwallis, dated 26th May, shows the same garrison still at the latter place with other British detachments, and, on Portsmouth being vacated in August, the whole of the Virginia force was assembled on the 22nd of that month at New York and Gloucester. Meanwhile Earl Cornwallis's army had attacked the Americans at Guildford Court House on the 15th March, and gained a victory.

After performing much harassing service, the troops under Earl Cornwallis took possession of York Town and Gloucester, where they were invested by the combined French and American forces on the 14th September.

They defended York Town until the works were destroyed by the enemy's batteries. The British lines were now falling to pieces, not a gun could be fired from them, and with a

* C.O. 5, Book 100.
† Stocqueler's History of the British Army.
‡ C.O. 5, Book 101.

shortage of ammunition and the reduced state of the troops from sickness and exhaustion, Earl Cornwallis, on the 17th October, considered it necessary to make proposals for a capitulation, the terms of which were adjusted the next day, and on the 19th, the posts of York Town and Gloucester surrendered to General Washington, and our ships of war, transports, &c., to the Comte de Gras as commander of the French fleet.

A famous "History of the United States" relates the capitulation as follows* :—

"At noon, on the 19th, one redoubt was delivered to the "French grenadiers and another to the American infantry. "At one o'clock the Gloster works were given up; at two "o'clock the British garrison at York Town marched out as "prisoners of war, with new uniforms but with colours cased, "and beating the old English march 'The world turned "'upside down.' Among the trophies were 75 brass and 69 "iron cannons; 18 German and six British regimental stan- "dards were given up."

The regular troops of France and America who obtained this important conquest consisted of about 7,000 French and 5,500 Americans, assisted by about 4,000 militia. Our killed and wounded amounted to 477, and 70 were taken prisoners in the redoubts on the 14th.

A Return of officers and privates, surrendered prisoners of war, on the 19th October, 1781, to the allied army under General Washington (taken from the original muster rolls), shows, including the Lieut.-General and his staff, a force of artillery, Guards, light infantry, and the following regiments: 17th, 23rd, 33rd, 43rd, 71st, 76th, 80th, two battalions Anspach, the Prince Hereditary's regiment, Regiment de Bose, Yagers, British Legion, Queen's Rangers, North Carolina Volunteers, Pioneers' Corps, engineers, commissariat and hospital depart ments; total, 7,247.

A Return of killed, wounded, and missing, from the 28th September to the 19th October, shows the 17th to have had one drummer killed and one sergeant and six rank and file wounded.

*E. Avery, Vol. VI., p. 318.

The regiment is shown as mustering, on the 19th October: One lieut.-colonel, three captains, eight lieutenants, four ensigns, one surgeon, nine sergeants, 13 drummers, 205 rank and file; total, 244.

The command of the army now devolved on Lieut.-General Sir H. Clinton, K.B.

A Distribution Return of the troops, dated 15th November, showed that the headquarters of the regiment had moved to Virginia.

The following General Order was published, dated Headquarters, Charlestown, 9th December, 1781* :—

Wound pensions granted by the King, who was graciously pleased, from an attentive regard to the distresses and services of his officers, to do so :—

a. Total loss of eye or limb.	A year's full pay and expenses for cure, if not performed at the King's charge.
b. If not equal to loss of limb.	Charge of cure only duly certified.
c. If killed in action.	Pension to widow and orphans; a year's full pay for widow; half for each child, under age and unmarried (posthumous included).
d. All persons dying of wounds within six months of battle to be deemed "killed in action."	

In April this year, a War Office circular was issued to all regiments of infantry in Great Britain directing that no man was to be enlisted under 5 feet 7 inches, and he was not to be over twenty-seven years of age.

1782

A "State of the Troops" in South Carolina in January, under Major-General the Honourable Leslie, shows a detachment of the regiment in camp at Charlestown, mustering three officers and 95 men, including two sergeants, one drummer, and four rank and file, prisoners with the rebels.†

* Major Brereton's Order Book.
† C.O. 5, Book 104.

Heavy sentences by general court-martial were, at this period, passed and approved for all breaches of discipline in the field.

Major Brereton's order book, referring to the war (in possession of the Officers' Mess, 1st Battalion), shows, under date of 20th January, 1782, that a sergeant and four privates of the British Legion had been tried for quitting their posts in search of plunder, and plundering the house of an American, and ill-treating his family, the sentence being that the sergeant and one private should suffer death, whilst the remaining three privates were to receive 1,000, 800, and 500 lashes respectively, all of which sentences were duly carried out.

For the crime of forgery, with intent to defraud a comrade, a soldier, tried by general court-martial at Charlestown on the 18th February, 1782, was sentenced to death, which was approved by Lieut.-General the Honourable Leslie, the confirming officer, and later remitted on account of previous good character.*

A General Order, dated, Losack Camp, 10th May, directed that, from the 1st January this year, all soldiers who had escaped from the enemy as prisoners of war, and had come into the British lines, were to receive one guinea each, and the same bounty was to be continued to all such who might in future rejoin their corps.

On the death of Lieut.-General the Honourable Robert Monckton on the 21st May, King George III. conferred the colonelcy of the regiment on Major-General George Morrison, from the 75th Regiment (afterwards disbanded), by commission dated the 29th of May, 1782.

By the 1st June, General Sir Guy Carleton had assumed command of the army in North America.

In August this year, orders were issued for the 17th Regiment to assume, in addition to its number, the title of "LEICESTERSHIRE," and to cultivate a connection with that county, which might at all times be useful towards recruiting.

On the 27th August, Major Brereton (who had been promoted from the 17th to a majority in the 64th) took a leading part in an action, in which the British regiments

* Major Brereton's Order Book.

engaged were detachments of the 17th, 64th, and 84th Regiments, besides provincials.

He had been sent to procure provisions on the borders of Georgia and South Carolina, for troops going to the West Indies, and a despatch from General Leslie, dated Charlestown, 8th September, says :—

" Major Brereton, who had landed a part of his force on " the Cambo, was immediately attacked by the advanced " guard of a detachment from General Green's army under " General Guest. The latter were repulsed with some loss, " and a howitzer taken, but their whole body coming up, " Major Brereton was obliged, within a few hours, to renew " the engagement with their force, which was now increased to " 300 infantry, about 60 cavalry, and two guns, to which he " could oppose only part of his detachment, about 140 men. " The howitzer taken by us, though disordered, was now of " some use, and the enemy were quickly dispersed with some " loss, one gun taken, and Colonel Laurens, with several " officers, in the number of their slain. All this was effected " by the loss on our side of one man killed and seven wounded. " I cannot withhold from the troops employed on this service, " that just praise their distinguished gallantry has so well " deserved; their comparative numbers and other circum- " stances of their situation will sufficiently evince the merit " of their conduct." This was the last action in the war, and Major Brereton, for the fifth time, received the thanks of the General commanding the troops in North America.

A " State of the Army," dated, 27th October, shows eight companies of the 17th in Virginia.*

1783

In the early part of the year the regiment was stationed at New York with a detachment at Kingsbridge; and a Disembarkation Return, dated New York, 7th January, shows that the detachment from Charlestown had joined there that day, mustering three officers, 74 men, 15 women, eight children.†

* C.O. 5, Book 107.
† C.O. 5, Book 108.

In the official books, there are two Returns of British, American, and German troops, who came into New York, from captivity, between the 8th May and 1st June this year. Amongst the British, are officers and men representing 24 regiments, including artillery and Guards, showing a total of five captains, 30 lieutenants, seven ensigns, two adjutants, three quartermasters, three surgeons, three surgeons' mates, 234 sergeants, 86 drummers, and 3,041 rank and file, of which one ensign, one sergeant, and 101 rank and file belonged to the 17th.*

Nova Scotia, 1783.

On the conclusion of peace with America, an order, dated, 17th August, directed the regiment to be held in readiness to move to Nova Scotia, whither it sailed on the 18th October, mustering 14 officers and 347 effective rank and file, and was stationed at Halifax.†

The War Office books show that contingent allowances to captains of cavalry and infantry were granted from the 25th December this year, and an annual allowance of £30 to each foot regiment for postage, stationery, guard rooms and store rooms taken together.‡

1784

In a War Office letter-book, a petition to the Secretary at War is shown from Captain Robert Clayton, 17th Regiment, dated, Halifax, N.S., May 25th, 1784, in which the petitioner sets forth that :—

He purchased his ensigncy in 1767, and his company in 1773, and since served in America in the following engagements, viz. :—

Long Island, Fort Washington, Brandywine, Germantown, Edgehill, Salem (with Colonel Mawhood), Monmouth, Horse Neck, Bedford, New England, and York Town, Virginia, and, with eighteen years' service to his credit, he represents the apparent hardship of so many captains junior to him in the service, being promoted to the brevet rank of major, submitting a list of the same.§

* C.O. 5, Book 109.
† C.O. 5, Book 111.
‡ W.O. 3, Book 26, p. 145.
§ W.O. 1, Book 3.

The reply to it was: "His Majesty is at present entirely "averse to the granting of any rank by brevet, however "strong the plea may appear on which the application for such "rank is forwarded."

The regiment was inspected on the 9th June this year, at Halifax, N.S., by Major-General John Campbell; on parade, 280 effective rank and file; "on command," 52 and 104 "wanting to complete."

The King gave orders this year, that for the future no leave was to be granted by any commander-in-chief or governor abroad for a longer period than six months. Should a prolongation be required. the application was to go before him.

The Muster Rolls (which at this period had to be declared before a Justice of the Peace and a Commissary of Musters) showed the regiment quartered at Halifax up to the 31st October, after which date it moved to Shelburne, N.S.

1785

A letter in the War Office books, dated, Halifax, N.S., 20th July, shows that Major-General Campbell was proceeding to Shelburne on the following day, to "review" the 17th, and from the muster rolls, it is clear that the regiment remained at Shelburne until required to embark for England in July the following year, from which it would appear, that the statement in "Cannon's History of the 17th Foot" (at top of page 29) as to its being quartered in Newfoundland, after its return from America, is incorrect.

CHAPTER VI.

HOME SERVICE, WEST INDIES, AND ST. DOMINGO.

1786–1798.

1786

To England, 1786.

HAVING been relieved from duty in North America, the regiment embarked in July for England, and a War Office letter, dated, 3rd August, directed it, on arrival at Portsmouth, to march to Guildford, and on the 18th and 20th September, it proceeded thence, in two divisions, by route march to Chatham.

1787

At the inspection at Chatham on the 4th May, by Major-General Sir G. Osborn, Bart., out of 345 effective rank and file, 150 were old soldiers, and the whole of the light company and more than half the grenadiers had served in the American War.*

A General Order, dated, 28th September, laid down, that recruits were to be taken from 16 to 35 years, and those for India were not to exceed thirty years of age.

On the 11th and 12th October, the regiment was directed to return to Guildford. Whilst on the march from Chatham, each of its two divisions received orders, on arrival at Bromley and Epsom respectively, to march to Hilsea Barracks, whence they were directed to proceed to Portsmouth and embark for Jersey, as soon as the vessels were ready for their reception. On arrival at Epsom, the Regiment was ordered to hand over a draft of 150 men to the 55th Regiment, each man receiving one and a half guineas bounty prior to embarkation.

1788

On the 22nd April, the regiment was ordered to move from Jersey to Dover and Deal (five companies to each), and being

* W.O. 4, Book 134, p. 27.

OFFICERS' SILVER BREASTPLATES.

(Exact sizes.)

1ST BATTALION. 1799 TO ABOUT 1815.

1776.

2ND BATTALION. 1799—1802,
WHEN THE BATTALION WAS DISBANDED.

much below its strength was directed to recruit to its full establishment.

At the inspection, by Lieut.-General Sir George Osborn, Bart., which took place at Dover Castle on the 26th May, it was composed so largely of recruits that no drill took place.

On the 1st October, the headquarter wing of the regiment was ordered to march to Chatham, and was followed, on the 3rd November, by the wing from Deal.

1789

A Route, dated 23rd September, directed the regiment to march from Chatham, five companies to Windsor, and five to Datchet, with four detachments at out-stations.

1790

An Order, dated 27th May, directed it to march in two divisions from Windsor to Petersfield and Alton, and remain until receiving notice of the 12th Regiment having embarked, when it was to proceed to Portsmouth.

Monthly Returns show that the headquarters were, in June, at Petersfield, in July, at Portsmouth, and from August to December at Hilsea.

In 1786, the year in which the regiment returned to England, a company of merchants residing in the East Indies had formed a settlement at Nootka Sound—a bay of the North Pacific Ocean, on the west coast of North America—with the view of obtaining furs. This settlement was seized by the Spaniards in 1789, and two ships were detained. To chastise this violation of British enterprise and liberty, a fleet was fitted out, and the 17th and 37th Regiments were embarked to serve as marines.*

An official list of quarters of the troops in Great Britain, dated, 7th August, 1790, shows the 12th, 17th, 29th, and 31st Regiments on board the fleet.

Embarkation Returns show that between the 19th June and 26th October this year, detachments of the 17th, totalling nine officers, 15 sergeants, 18 corporals, 8 drummers, and

* Adjt.-Genl.'s Letter to Lord Dorchester, dated 7th July, 1790.

349 privates had embarked, to serve as marines, on the following ships, viz., "Princess Royal," "Edgar," "Colossus," "Saturn," "Alfred," "Duke," and "Vengeance."

The difficulties above mentioned, having been settled without hostilities, the detachments of the regiment disembarked at Gosport in November, and on the 16th, the headquarters were ordered to march to Salisbury, with detachments at Fiskerton, Harnham and Milford.

1791

On the 12th February, an order was issued for the regiment to march in two divisions, from Salisbury to Plymouth.

1792

A Royal Warrant, dated, 14th March, placed the establishment of the regiment at ten companies, each to consist of two sergeants, three corporals, one drummer, 30 private men, two fifers to the grenadier company, with the usual commissioned and staff officers, "and two sergeants in excess of those "mentioned and no more," with effect from the 25th June; and a letter from the Secretary at War of the same date directed the establishment to be completed by supernumeraries from the 25th Foot, each transfer receiving one and a half guineas bounty.

Ireland, 1792.

In June, the regiment received orders to proceed to Ireland, its duties at Plymouth being taken by marines; but no transports being available, it did not embark until October, in three ships named "Friendship," "Helmsley," and "Constant Trader" (total 286 privates), and a Disembarkation Return shows that it landed at Monkstown on the 10th of that month and marched to Kilkenny.

A War Office letter, dated 25th July, authorised the officer commanding to have the sum of £6 16s. 8d. paid to each of such men of the regiment, from whom it had been stopped on account of provisions whilst prisoners of war in North America.*

An order, dated 3rd September, directed that a stoppage of 3d. a day for provisions was to be made from all ranks whilst on board ship.

* W.O. 4, Book 144, p. 380.

Lieut.-General Morrison was removed to the 4th Foot in 1792, and was succeeded in the colonelcy of the 17th Regiment by Major-General George Garth, from lieut.-colonel in the 1st Foot-guards.

1793

In January, the regiment marched from Kilkenny to Dublin. From the 26th July, 400 men were ordered to be raised by each regiment serving in Ireland, by which two captains, two lieutenants and two ensigns were added, and from the 1st December, 11th and 12th companies were added to each of these battalions, the additional not to be flank companies, and also a second lieut.-colonel, and a second major to each battalion, special terms of promotion for officers being stated.*

On the 13th and 14th August, the regiment marched to Mullingar.

The Monthly Return of the 1st September shows it at Granard, and that of 1st November, at Galway.

The Government having decided to deliver the French West India Islands from the hands of the republicans (who had added to their other atrocities the decapitation of their sovereign), an expedition was sent this year to the West Indies, under Sir Charles Grey and Admiral Sir John Jervis.

A letter from Lord Westmorland to the Honourable Henry Dundas (Secretary of State), dated, 23rd September, stated that, on the 20th instant, he had given orders for the flank companies of the following 14 regiments, viz., 8th, 12th, 17th, 22nd, 23rd, 31st, 33rd, 34th, 35th, 38th, 40th, 41st, 44th, 55th, to be completed to four sergeants and 100 rank and file each, the same when completed to march to Cork to embark; also 10 supernumerary rank and file per company, in order to embark complete. Also this force to be properly officered, and he added: "I shall divide it into four battalions and appoint field officers, with a proper staff to each."

The 1st Battalion Grenadiers, commanded by Lieut.-Colonel Sir C. Gordon, embarked at Monkstown on the 13th November, and consisted of the grenadier companies of the 8th, 12th, 17th, 22nd, 23rd, 31st and 41st Regiments.

* W.O. 8, Book 8, p. 391, and "Anthologia Hibernica, Domestic," for Dec., 1793, folio 470.

(The 17th, under Captain Hamilton, mustered one captain, two lieutenants, four sergeants, four corporals, two drummers and fifers, 65 privates, six women and three children.)

The 2nd Battalion Grenadiers, commanded by Lieut.-Colonel Cradock, embarked with the 1st Battalion, and consisted of the grenadier companies of the 33rd, 34th, 35th, 38th, 40th, 44th and 55th Regiments.

The 1st Battalion Light Infantry, under Brevet-Major Ross, embarked at the same time and place, and consisted of the light companies of the 8th, 12th, 17th, 22nd, 23rd, 31st and 35th Regiments.

(The 17th, under Captain Stovin, mustered one captain, two lieutenants, one quartermaster, four sergeants, four corporals, two drummers and fifers, 66 privates, six women and one child.)

The 2nd Battalion Light Infantry embarked at the Cove of Cork on the same date, and consisted of the light companies of the 33rd, 34th, 38th, 40th, 41st, 44th and 55th Regiments.*

1794

The above troops, which left Ireland with Sir Charles Grey, arrived at Barbados on the 6th January.

The operations commenced with an attack on the island of Martinique, which had been captured by the British in 1762, and ceded to the French by treaty the following year.

The entire armament of 19 ships of war, escorting 7,000 soldiers, sailed from Barbados on the 3rd February, in three divisions for Martinique, and amongst the troops were the four battalions of grenadier and light companies from Ireland.

By the 8th February, the force had arrived at three different points of the island, and after some sharp fighting, the island of Martinique was taken by the 22nd March.

From Martinique, the battalions of grenadiers, under command of Prince Edward (afterwards Duke of Kent), with some artillery and three other regiments, embarked on the 30th March, for St. Lucia, and its conquest being promptly effected with no loss on the British side, this expedition returned to Martinique on the 5th April.

* Irish Government Correspondence Book, No. 136.

The flank companies of the 17th were afterwards employed in the reduction of Guadaloupe, when a determined resistance was made, but the island was captured by the 20th April.

The Commander-in-Chief, in his despatch home on the capture of Guadaloupe, stated that he could not find words to convey an adequate idea, or to express the high sense he entertained of the extraordinary merit evinced by the officers and soldiers in this service.

France did not view with indifference the loss of these valuable possessions, and in June, a French armament, under command of the republican commander, Victor Hugues, was sent to Guadaloupe for the recovery of that island. It so happened, that the small French fleet, with about 1,500 troops on board, escaped the notice of the English cruisers, arriving at Point à Pitre, Guadaloupe, on the 3rd June.

The British troops being now much incapacitated, owing chiefly to yellow fever, the French commander took Point à Pitre by storm, and having succeeded in driving out the feeble garrison from Grande Terre, the English commander took up a new position at Basse Terre. The command of the troops at Basse Terre had devolved on Lieut.-Colonel Graham, 21st Fusiliers, who took up a final position at Beville camp, which he defended with the utmost gallantry and spirit until the 6th October, when finding his provisions nearly exhausted, and that he was cut off from all communication with the shipping, and without hope of relief, he was obliged to surrender, his force being reduced to 125 rank and file fit for duty.*

In Sir Charles Grey's Return of killed, wounded and missing at Grande Terre, Guadaloupe, from the 10th June to 3rd July, 1794, Lieutenant Auchmuty, grenadier company of the 17th, is shown as wounded.

The prisoners taken by the French were, according to the stipulation, to have been allowed to go on board the English ships, but the agreement was not complied with, and they were kept prisoners for over a year, during which many died owing to the severity of their confinement.

* "London Gazette," Dec. 13th, 1794.

The troops that capitulated were the flank companies from Ireland, of the 8th, 12th, 17th, 31st, 33rd, 34th, 38th, 40th, and 55th Regiments, the 39th, 43rd and 65th Regiments, and three companies of the 56th. (Sir C. Grey's despatch, Bulletin, 1794, p. 357, and W.O. 1, Book 83, P.R.O., London.)

The Monthly Return of the regiment, dated, 1st July, shows the headquarters at Galway. Without its flank companies, it embarked at Cork on the 10th August, in the ships "Kenvicus," "Catharine," "Lavinia," "Nancy," "Lovely Martha," "Prince of Wales," "Helena," "Jane and Ann," and "Sally," and was directed to disembark at Bristol and to march, five companies to Frome and five to Warminster and Heytesbury, and thence (20th August) to Netley Camp.

On the 29th August, orders were received to march to Southampton and embark on Sunday, 31st August, no destination being stated.

An Embarkation Return, dated Southampton, 2nd September, shows its strength as 690 of all ranks, also 125 women and 94 children, but no destination is stated.

A War Office letter, dated, 26th September, fixed the establishment of the regiment at ten battalion companies (each consisting of three sergeants, three corporals, two drummers, and 57 private men), with a sergeant-major and quartermaster-sergeant, and also two flank companies, each consisting of four sergeants, four corporals, two drummers, and 66 private men, with two fifers to the grenadier company. Authorised to commence from the 10th August.*

Sir James Vaughan succeeded to the Governorship of Barbados in November, and in a despatch to the Secretary of State, dated, Martinico, 21st December, 1794, he wrote: "We "approach the end of December, and only one transport is "yet arrived, with 100 of the 17th Regiment." He further complained, after the disaster at Guadaloupe, that the reinforcements expected, viz., 17th Regiment, 800, 31st and 34th, each 700, from England, besides three regiments from Gibraltar, were not nearly sufficient to retake Guadaloupe.

A Return of the disposition of troops in the Charibee Islands for the month of December shows a part of the 17th "detained" for duty in Dominica, Saints, and Mary Galante.

Beyond the 100 men of the regiment sent this year to the West Indies, in addition to the flank companies already there, the headquarters and remaining companies were on board transports in Plymouth Harbour, awaiting sailing orders, from the 1st September to early in February, 1795, as the Monthly

* W.O. 4, Book 154.

Return, dated, 1st November, 1794, shows the regiment in Cawsand Bay (off the Cornish coast), whilst that of the 1st December was signed in Plymouth Harbour, and that of the 1st February following, shows the headquarters in Plymouth Sound.

1795

On the 11th February, the regiment was directed, on arrival off Portsmouth, to disembark and be quartered "in "such unoccupied barracks in the Isle of Wight as may be "considered most expedient."

This was, however, changed to Plymouth, where it was quartered in barracks until August 25th, when it embarked for Cork in the ships "Comet" and "Flora,"* and was quartered at Spike Island, where it was sent merely to receive drafts, to complete it to the establishment, and was not placed on the Irish establishment, prior to embarking for foreign service, the authorised establishment now being: one colonel, two lieut.-colonels, two majors, nine captains, one capt.-lieutenant, 25 lieutenants, 10 ensigns, 60 sergeants, 26 drummers and fifers, and 100 rank and file per company.

A letter to General Garth (Colonel of the regiment), on the 27th August, ordered its completion to 1,000 rank and file.

Lieut.-Colonel Frederick Augustus Wetherall, who had been gazetted to the 17th as Ensign in 1775 (and in 1840 became its Colonel), now had the misfortune to be made a prisoner of war. He had fought with the regiment throughout the American War, and after exchanging to the 11th Regiment had, in May this year, been appointed Lieut.-Colonel in a corps which afterwards became the 3rd West India Regiment. (See Biographies of Colonels.) Whilst serving on the staff at St. Domingo, he had been detailed to convey despatches to Sir R. Abercromby at Barbados, and having been taken prisoner by a French frigate, and wounded in action, remained in confinement at Guadaloupe for upwards of nine months.

Whilst closely confined in a dungeon, in irons, without any description of bedding, his only clothing was a shirt and pair of trousers, whilst his daily allowance of food consisted of three biscuits and a quart of water.†

* Embarkation Return. † Hart's Army List, 1841.

In connection with this, the following humane act is reported: While Colonel Wetherall was a prisoner under such distressing circumstances, some men of a detachment, consisting principally of the 32nd Regiment (with which he had served at Gibraltar), were taken prisoners on their passage from that garrison to Barbados, and all were brought to Guadaloupe. On their hearing of the severe and inhuman treatment to which Colonel Wetherall (by order of the revolutionary governor of the island) was being subjected, they made a collection amongst themselves of eleven guineas, and forwarded it to him, concealed in a small loaf of bread, through the medium of a negro employed in selling provisions, with a note from a sergeant of the 32nd, requesting the Colonel's acceptance of the money, in the name of that corps, and of other unfortunate companions in captivity, as a small token of esteem, and in the hope of its affording him some relief and comfort under the sufferings and cruel treatment they understood he was undergoing.

This noble act was mentioned to the Commander-in-Chief, the Duke of York, who conferred an ensigncy on the sergeant.*

A Muster Roll of the grenadier company of the regiment, under Brevet-Major Hamilton (taken in the West Indies for the period ending 24th June, 1795), shows it, much reduced by death, to number only three sergeants, three corporals, and eight privates.

A Return of the Army, in the Windward, Leeward, and Charibee Islands, dated July 7th, 1795, shows detachments of the regiment distributed as follows: At Barbados, one officer and eight rank and file; at Dominica, four officers and 67 rank and file; at Saints, five rank and file.

Three more companies of the regiment had gone to the West Indies in April this year, as the Monthly Returns from May to August show four companies there, besides the flank companies.

St. Domingo, 1795.

An Embarkation Return, dated, Cork, 10th December, shows that the headquarters of the regiment embarked on that date for St. Domingo, in the following transports, viz., "Maria," "Grampus," "Cornet," "John and Sarah," and

* By the courtesy of Captain F. A. Wetherall, R.N., retired.

"Blessing." Its embarkation strength was: one major, six captains, 12 lieutenants, seven ensigns, one quartermaster, two surgeons, two surgeons' mates, 36 sergeants, 33 corporals, 19 drummers, 555 privates; showing 10 officers, including the adjutant, with four companies in the West Indies.

1796

A contest was now being carried on between British troops and the republican forces in St. Domingo, and, in the first week of May, a part of the long-expected reinforcements arrived from England, both British and foreign troops, comprising the 17th, 32nd, 56th, 67th Regiments, detachments of the 39th, and two other regiments shortly afterwards disbanded, the foreign troops being Lewes's Foot, York Chasseurs, and Rohan Hussars.

An extract from a statement by the Privy Council of St. Domingo, dated, 14th April, 1796, describes the colony as having been "for many years the richest European possession "in the two Indies, whilst the vast extent of its coasts, the "beauty of its harbours, the fertility of its soil, the industrious "activity of its inhabitants, had raised its culture and com- "merce, to such a degree of splendour, it is to be feared, it will "never again attain."*

The following accounts of skirmishes with republicans are published from the despatches of Major-General Bowyer and Brigadier-General Churchill:—

"From Major-General Henry Bowyer,

"Government House, Jeremie, September 3rd, 1796.

"The posts on the extremities of the Dependency, both "east and west, being attacked in force on the same day, I "ordered, on the 8th August, Lieut. Bradshaw, with 22 of the "13th Light Dragoons, mounted, to march from Du Centre, "and Captain Whitby, two subalterns, and 60 privates of the "17th, with non-commissioned officers in proportion, embarked "the same evening for Cazinites, which they reached without "opposition. On Captain Whitby's arrival at Du Centre, he "detached to Post Raimond 20 men of the 17th, under Lieut. "Gilman, who immediately occupied the Block House with his "detachment, and a large party of chasseurs.

* W.O. 1, Book 64.

"On the 11th, the enemy appeared before Post Raimond, "and, after keeping up a very heavy fire on the Block House "with little effect, they stormed it in considerable force four "times, and were each time repulsed with great loss; and their "chief killed. On the 12th, they continued, with as ineffectual "a fire as before, when Lieut. Gilman, with the 17th, and some "chasseurs, made a successful sortie, driving the enemy into "the woods, who left 16 white and 47 blacks dead on the spot, "many more of their dead and wounded being afterwards "found in the woods.

"I am happy to report, that, in this gallant affair, the 17th "had only two privates wounded; the chasseurs had one officer "and three chasseurs killed and 14 wounded; and the enemy's "loss is believed to be considerable. Lieut. Gilman's intrepidity "and coolness appear to me so praiseworthy, that I strongly "recommend him for promotion.*

"On the 11th August, General Rigaud, with a view to "raising the siege of Irois, appeared with three or four thousand "brigands before it, and sent a summons to Captain Beamish, "to surrender the fort to the republicans of France, who very "properly replied, that he would defend it to the last extremity.

"The next day, the republicans opened fire on the fort, "and, on my hearing of Irois being invested, I ordered 100 "privates, (with officers and non-commissioned officers in "proportion), of the 17th, under Lieut.-Colonel Hooke, to "embark for that place. Colonel Hooke took command of "the fort of Irois, in which station he has rendered very meri- "torious service.

"Finding that the enemy had advanced a considerable "distance, between Irois and Lance Enos, I determined, with "what force I had, to attack them on Mount Gautier, where "they had established themselves to the number of twelve to "fifteen hundred, and had attempted to surprise Post D'Islet, "from which they had been repulsed with very material loss.

"On my arrival at Lance Enos, (16th), a plan was agreed "on, to march in three columns on the 19th, so as to arrive "before Mount Gautier at daybreak. The dispositions of the "attack by the three columns had been laid down, and the

* Lieut. Gilman was promoted Captain into the 56th Regiment ("London Gazette," 20th Dec., 1796).

" execution of it depended on the calculation of time and
" place, which, I am sorry to say, was not attended to.

" I was in command of the centre column, and had with me
" about 90 of the 17th, the others having been chiefly negroes,
" clothed and armed two days before, wholly unacquainted with
" the use of arms, and, when the firing began, lay down to a man,
" and nothing would prevail on them to move forward, though
" the 17th were in front and set them a good example. Observ-
" ing, on our approach, that a few men of the 17th were killed
" and wounded, and that firing at a distance was of no avail, I
" determined to carry the hill by assault, and had formed the
" 17th for that purpose, ordering the 13th Light Dragoons to
" dismount, and was endeavouring to rally the negroes, who
" had been thrown into some confusion, when I received a shot
" in my left breast, which caused me to fall from my horse.
" After that I knew nothing except by report, and was sorry
" to hear we were obliged to retreat with the loss of the
" three-pounder we had with us. As to Colonel Sevre's column
" (who, attacking the rear, was to cut off the enemy's retreat),
" we never saw or heard anything, the enemy having advanced
" an ambuscade two miles nearer Irois, which he attacked,
" and was repulsed, he himself being mortally wounded.

" Fortunately these little checks did not affect the safety
" of the fort. On the contrary, the enemy, who must have
" suffered more than ourselves, evacuated Gautier and retired to
" General Rigaud, on the other side of Irois, and on the 29th,
" Colonel Hooke wrote me that he had raised the siege entirely.

" I have not yet been able to procure exact returns of the
" killed and wounded, but the 17th Regiment had about seven
" killed and 14 or 15 wounded, some dangerously, in my affair
" of the 19th ultimo. Lieut.-Colonel Hooke has not yet made
" his report, but I don't believe more than two were killed at
" Irois, and three or four wounded, so that the enemy's shot
" and shell had little effect.

" I propose that Jeremie shall be a garrison of conva-
" lescents from the army, which may enable me to keep most
" of the 17th in the country.

" Major Hertzog, of the 17th, died a few days ago on board
" the 'Iris' at Irois, and Lieutenant Hope died also there
" of fever."

As an instance of the sufferings from the climate, the following is an extract from a letter, from General Gordon Forbes to the War Office, dated, Port-au-Prince, St. Domingo, 25th October 1796 :—

" The 82nd Regiment is almost reduced to nothing, and " as most of the officers of the 17th are sick or dead (in con- " sequence of the severe service they underwent, and the " effects of the dreadful climate they experienced at Irois, " when attacked by Rigaud), I have sent the small remains " of the former regiment to Jeremie, in order, in some " measure, to enable General Bowyer to replace the losses " he has sustained in that quarter." *

The Records of the 41st (Welsh) Regiment (page 40), state that, when the 41st embarked, in the West Indies, for Jamaica, on the 5th July, 1796, the regiment drafted its private soldiers to the 17th Foot, and landed in England with three men only besides the commissioned and non-commissioned ranks.

Lieut.-Colonel Hooke, commanding the 17th, died at Irois in October.

1797

A despatch from Brigadier-General Churchill, dated, Jeremie, April 30th, states :—

" On the night of the 20th April, Rigaud thought it a " favourable moment to make a second attempt on Irois. " Accordingly he collected his best troops, amounting to " 1,200, and at 12 o'clock they attempted to storm the fort, " in which were only, at that time, 25 of the 17th Regiment, " with officers, commanded by Lieut. Talbot, 82nd, and about " 20 colonial artillery. The attack was one of the most for- " midable and determined, the besiegers returning to the " charge on three separate occasions, with such increased " vigour that many of them were killed in the fort ; but, to " the immortal honour of its brave defenders, they were " repulsed with equal courage and intrepidity, which gave " time to Colonel Dugress (with 350 of Prince Edward's Black " Chasseurs), to gain the fort from the Bourg below, whence " they were obliged to cut their way. This reinforcement

* W.O. 1, Book 65.

REGIMENTAL MEDAL, 1810.
Presented in 1811.

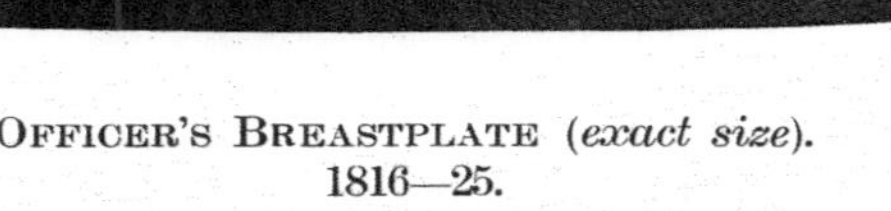

OFFICER'S BREASTPLATE (*exact size*).
1816—25.

REGIMENTAL SCHOOL MEDAL,
1816.

"saved the place, as it would have been impossible for the "besieged to have withstood the persevering and repeated "attacks of a most desperate enemy, which continued until "morning, when they retired, leaving the fort surrounded "by their dead, to a higher ground, where they made a stand, "in spite of a sortie that was immediately made with some "advantage.

"Here they continued until the 22nd instant, when they "made an incursion into the interior of our cordon, burnt the "Bourg of Dame Marie, and attacked the fort of Lilet, "whence they were driven with great loss."

The despatch continues with two minor engagements, and adds:—

"The loss of the enemy in all these various attacks is "generally estimated at 1,000 men, and cannot be less than "800. Our loss, consisting of three English and colonials, "is trifling."*

In May, this year, a warrant was issued revising the pay and allowances of the rank and file, the pay of the sergeant being fixed at 1s. 6¾d.; of a corporal, 1s. 2¼d.; of a drummer and fifer, 1s. 1¾d., and of a private man, at 1s.

1798

A return of the British troops, present and fit for duty in St. Domingo, on the 1st July, shows the strength of the regiment as 239 effective rank and file.†

In addition to the deaths of Lieut.-Colonel Hooke, and Major Hertzog, (both of the regiment), a great number of men had died during its stay at St. Domingo, and it is much to be regretted that no complete statement of the casualties is procurable.

On evacuating the island, in September, instructions were received‡ to send all the British troops to Jamaica, on their way to England.

* W.O. 1, Book 67.
† W.O. 1, Book 68.
‡ W.O. 1, Book 70.

CHAPTER VII.

HOME SERVICE, HELDER EXPEDITION, ENGLAND AND MINORCA.

1799–1801.

1799

England, Jan., 1799

A DISEMBARKATION RETURN shows that the regiment arrived home on the 25th January, landing at Deptford, and a War Office letter, dated 26th January, directed that after detaching the invalids to Chatham, the remainder of the regiment was to disembark and march to Leicester, where, according to the Monthly Returns, it was quartered from March to May, and marched in June to Norwich.

The following regulations relative to the supply of bread to troops in camp were issued, dated Whitehall, 5th June :—

" A Well Baked Loaf, weighing six pounds, made of " good English Wheat, out of which the first or Coarse Bran " hath been taken by means of an Eight Shilling Cloth, will " be delivered to each Soldier as his Allowance of Bread for " four days.

" For this Loaf the Soldier is to pay four pence, five-eighths, " which is to be stopped in the hands of the officers paying " Troops or Companies, till the Settlement of the Account, " by the Regimental Quarter-Masters, in the manner after- " mentioned.

" Servants and Batmen, as also the Women, are permitted " to receive Bread at the same price."*

A regiment of 10 companies, 70 men in each, was to have 160 tents, 160 tin kettles and bags, 160 wood hatchets, 12 bell tents, 12 camp colours, 20 drum cases, 10 powder bags, 792 water bottles, haversacks, and knapsacks.

* W.O. 3, Book 19.

A favourable opportunity appearing to present itself, for rescuing the United Provinces of Holland from the degrading tyranny of the French Republic, an agreement was entered into between Great Britain and Russia, to restore the Dutch to their independence, under the authority of the Prince of Orange,* Russia finding 13,000 and England 17,000 men.

But the spring of 1799 found England with a seriously weakened army. An arduous campaign in the West Indies, the mortality incidental to that colonial service, and the many serious affairs in which the army had been engaged in Europe, had caused such a reduction in strength that no ordinary mode of recruiting could supply the deficiency. There were not in England at this time more than 12,000 infantry, comprising battalions of sufficient strength that could be spared for foreign service. Many battalions were quite skeleton in point of numbers, and a large number were worn out after arduous service in the East and West Indies.

To meet the difficulty caused by this enormous shortage of recruits, the British Parliament passed an Act (12th July, 1799), with a view to completing regiments depleted by fever, disease, and war losses, by volunteers from the embodied militia.

The militia responded with alacrity, and between the 18th July and the 15th November no less than 1,558 joined the 17th Regiment.

The large number of volunteers received induced the authorities to form a second battalion of the regiment, and in a gazette, published on the 10th of August, 1799, the following officers were posted to the new battalion :—

Major-General Eyre Coote, from the 70th Foot, to be Colonel Commandant (13th August).

To be Lieut.-Colonels :—Major Wm. Francis Farquhar (29th Foot), Major David Latimer Tinling.

To be Majors :—Bt. Lieut.-Colonel Robert Wood (15th Foot), Captain Peter Cockburne, Captain St. John Fancourt (56th Foot).

To be Captains of Companies :—Captain Michael John Grace, Captain Wm. Maxwell (late 93rd Foot), Captain P. H.

* W.O. 1, Book 179.

Nichol (h.p., late 122nd Foot), Captain Rollo Gillespie (20th Foot), Captain Wm. Knight, Captain Chas. Nowell (R. Fusiliers).

To be Capt.-Lieutenants:—Lieut. Wm. Baillie (72nd Foot), Lieut. Evan MacPherson (92nd Foot).

To be Lieutenants:—Ensign Francis Boynton, Ensign G. W. T. Tinling, Ensign Jas. Walker, Ensign John Steel, Ensign W. Metcalf, Ensign James Vallance, Ensign Chas. Wilson, Ensign Wm. White, Ensign Wm. Pearson.

To be Ensigns:—Lieutenant J. Spence (3rd Lincoln Militia), Lieutenant C. F. Champion (S. Gloster Militia), Lieutenant J. Thompson (3rd Norfolk Militia), Lieutenant G. Burridge (W. Somerset Militia), Lieutenant J. Winton (S. Essex Militia), Lieutenant Brownson (E. Essex Militia), Lieutenant Sinclair (1st Northumberland Militia), Lieut. S. Bromley (S. Essex Militia), Lieutenant Thos. Walker Chambers (N. Middlesex Militia), Ensign Tyndale (E. Norfolk Militia), Ensign Richard Midgeley (E. Norfolk Militia), Ensign P. H. Mahoney (? Money) (E. Norfolk Militia), Ensign Ed. Lenn (Westminster Vols.), Ensign Jas. Hitchens (Oxford Militia), Ensign A. Macdonald (3rd Surrey Militia).

Lieutenant W. Kirk to be Adjutant.

Sergeant Speedy to be Quartermaster.

J. Bullock to be Surgeon.

The 1st Battalion had moved to Norwich in June, where the 2nd Battalion was also concentrated on formation, and, on the 5th August, orders were received for both battalions to march to Canterbury, where they were encamped at Barham Downs, prior to embarking for active service in Holland.

Helder Expedition, 1799.

The fleet, commanded by Admiral Mitchell, conveying the advance guard of the expedition, sailed from the Downs on the 13th August for the Helder, carrying 12,000 men, under Lieut.-General Sir Ralph Abercromby, K.B., but was delayed by the roughest of weather, so that, although being in sight of the shore for upwards of six days, it failed to reach its proposed anchorage until the 26th of the month, thus enabling General Daendels, the Dutch commander, to assemble his troops, some 10,000, to oppose the threatened landing.

At daybreak next morning our troops, under the command of Lieut.-General Sir James Pulteney (second in command), landed in flat-bottomed boats under cover of the fire of the guns of the whole fleet, which however did no serious damage to the enemy, as Daendels had hidden his men from direct fire in the folds of the sandhills. He, however, retired about 3 p.m., after having lost 1,400 men in vain endeavours to dislodge the British right, now established in the sandhills, and the British landing was completed, with a loss of 454 killed and wounded.

The arrival of the 1st Division of Russian troops depended very much on wind and weather, but it was expected to reach Yarmouth between the 15th and 25th August, the remainder of the Russian contingent being expected between the 1st and 16th September; the 3rd Division, however, did not arrive until the 24th September.

On the 28th August, reinforcements arrived from the encampment at Barham Downs, which had embarked at Deal, and consisted of two brigades 5,000 strong, Major-General Don's brigade comprising two battalions of the 17th and two battalions 40th, the 20th and 63rd being in the Earl of Cavan's brigade. They disembarked on the same date under the batteries of the Helder, and at once joined Sir Ralph Abercromby.

An Embarkation Return shows the total rank and file of each battalion of the 17th embarking as: 1st Battalion 645, 2nd Battalion 641, with 155 and 159 respectively, "wanting to complete."

The same day, a detached force under Major-General Sir John Moore had taken possession of the Helder, and seized its arsenal and naval stores, and two days later, Admiral Mitchell forced the Dutch Admiral to surrender, and hoist the Orange flag, so that within a week the English object of the expedition had been achieved with little loss.

On the 1st September, General Abercromby occupied the position of the Zuyper, and here waited, knowing the Russian army was on its way, and that the Duke of York with more reinforcements would arrive in a few days.

General Daendels, on his defeat on the 27th August, had retired towards Alkmaar, where he was reinforced on the 2nd September by the French Commander-in-Chief in Holland, and on the 10th, the combined Franco-Dutch force, 20,000 strong, attacked the Zuyper position, and after a desperate fight, in which they lost 2,000, they were defeated and retired to their original position. The English loss was slight.

The casualties of the 1st Battalion were two rank and file killed, and of the 2nd Battalion, two rank and file killed and 18 wounded.

On the 13th September, the Duke of York, who had been appointed Commander-in-Chief of the allied forces, arrived and took command, and the Russian troops having landed the previous day, he decided to at once take the initiative and attack the enemy.

The allied forces now mustered 35,000 men (of which 18,000 were British), and on the 19th, the army moved forward in four principal columns.

The following extract is from the Duke of York's despatch, dated Schrager Brug, September 20th, 1799.

" The left column under Sir R. Abercromby, destined " to turn the enemy's right on the Zuyder Zee, marched at 6 p.m. " on the 18th. Of the columns on the right, the 1st com- " manded by Lieut.-General Hermann, consisted of the 7th " Light Dragoons and 12 battalions of Russians and Major- " General Manners' brigade.

" The 2nd column was commanded by Sir Henry Dundas, " and the 3rd by Sir James Pulteney, and in the latter were " included Major-General Don's (in which both battalions " of the 17th were serving), and Coote's brigades.

" Lieut.-General A. Hermann's column commenced the " attack, which was conducted with the greatest spirit " and gallantry at 3.30 a.m., and by 8, had succeeded in " so grand a degree as to be in possession of Bergen. " In the wooded country which surrounds this village " the principal force of the enemy was placed, and the " Russian troops advancing with an impetuosity which " over-looked the resistance with which they were to " meet, had not retained that order which was necessary to " preserve the advantages they had gained. They were in " consequence, after a most vigorous resistance, obliged to " retire from Bergen, where, I am much concerned to state, " Lieut.-Generals Hermann and Ichertchekoff were made " prisoners, the latter dangerously wounded, and fell back

" upon Schozel, which village they were also forced to abandon,
" but which was immediately retaken by Major-General
" Manners' brigade, notwithstanding the very heavy fire of
" the enemy. Here, this brigade was immediately reinforced
" by two battalions of Russians, by Major-General D'Oyley's
" Brigade of Guards, and the regiment under command of H.H.
" Prince William. The action was renewed by the troops for a
" considerable time with success, but the entire want of
" ammunition on the part of the Russians, and the exhausted
" state of the whole corps engaged in that particular situa-
" tion, obliged them to retire, which they did in good order
" upon Pitten and Zyper Sluys.

" As soon as it was sufficiently light, the attack upon
" the village of Walmen Naysen was made by Lieut.-General
" Dundas.

" Three battalions of Russians very gallantly stormed
" the village on its left flank, while at the same time it was
" enclosed on the right by the 1st Regiment of Guards. The
" remainder of Lieut.-General Dundas's column, after taking
" possession of Walmen Naysen, was joined by another
" regiment, and marched against Schoreldam, which place
" they maintained under a very heavy and galling fire, until
" the troops engaged on the right had retired at the conclusion
" of the action.

" The column under Lieut.-General Sir James Pulteney
" proceeded to its object of attack at the time appointed,
" and carried by storm the principal post of Oudt Carspel,
" at the head of Lange Dyke. This point was defended
" by the chief force of the Batavian army under General
" Daendels.

" The circumstances, however, which occurred on the
" right, rendered it impossible to profit by this brilliant
" exploit, and made it necessary to withdraw Lieut.-General
" Sir James Pulteney's column from the position he had
" taken within a short distance of Alkmaar.

" The same circumstances led to recalling the corps
" under Lieut.-General Sir R. Abercromby, who had pro-
" ceeded to Hoorn, which city he had taken possession of with
" its garrison. The whole army had therefore reoccupied its
" former position.

"The gallantry displayed by the troops engaged, the "spirit with which they overcame every obstacle which "nature and art opposed to them, and the cheerfulness "with which they maintained the fatigues of an action which "lasted without intermission from 3.30 a.m. until 5 p.m., are "beyond my powers to describe or extol."

His Royal Highness, in his despatch, then expresses his obligations to several Generals and members of his staff, and adds: "Not having received returns of the loss sustained "by the Russian troops, I can only observe that I understand "their loss in killed, wounded, and missing, amounts to nearly "1,500 men."*

Lieut.-General Sir James Pulteney's despatch, dated 20th September, 1799, says:

"The two brigades under my command were at their "destinations, according to the orders of H.R.H. the "Commander-in-Chief, at daybreak on the 19th instant, "that of Major-General Coote at Verlaght, and that of "Major-General Don in front of Drikshorne. I accompanied "the latter.

"The 2nd Battalion 17th Regiment was detached to "the road on the left, and I proceeded with the remainder "of the brigade along the dyke leading to Vost Cappel. The "enemy's outposts were driven in by the light infantry "without much resistance. The head of the village was "strongly fortified with a double row of entrenchments, "containing 8 or 10 guns, and a canal, a short distance "from this entrenchment, the bridge over which the enemy "had broken down, but there was still room to pass over it "singly. In front of the canal, and very close to the en-"trenchments, was a dyke, which afforded a good position, "and appeared absolutely necessary to take. I therefore "immediately directed part of the 1st Battalion 17th Regiment, "the 1st Battalion 40th, and three companies of light infantry, "to pass the canal and take that position, putting them "under command of Colonel Spencer of the 40th. The enemy "made several attempts to dislodge these troops, but were "constantly repulsed. They had also an advanced work in "a meadow, in front of the 2nd Battalion 17th, under cover

* W.O. 1, Book 179.

"of which they made several fruitless attempts to drive back "that battalion and to turn our left flank. The object "now was to obtain communication with Major-General "Coote, who had been endeavouring to force the enemy on "the other side of the village, but his brigade not coming "up, General Pulteney resolved to carry the village by an "attack from the position of the dyke. The enemy made "a vigorous attempt to turn the right flank of Colonel "Spencer's position, and renewing their attack, were, on each "occasion, repulsed with great loss, being finally driven "through the whole chain of villages which form the Lange "Dyke, the British taking upwards of 700 prisoners, 14 "guns, and a great quantity of ammunition."

Major-General Coote, in his report, dated Schragen, 20th September, 1799, describes the position he had taken up with his brigade, in keeping up a constant cannonade on the village of Oudt Carspel and battery, and adds: "Perceiving that "the 17th and 40th Regiments had gained a considerable "advantage in their front, and the enemy retiring before "them to the village, I lost no time to favour their advance, "and pushed the light companies of the Queen's and 29th "forward to the battery which the enemy evacuated. I "immediately directed those corps to cross the canal in a small "boat, and joined the 17th and 40th Regiments under Colonel "Spencer. When we crossed the canal, we made 8 officers "and 155 men prisoners, whom I sent to join headquarters, "and at about 12 midnight I arrived at Schragen with my "brigade."*

The regimental losses in this engagement were :—

1st Battalion, killed, 6 rank and file; wounded, Majors Grey and Cockburne, Captains Grace and Knight, Lieut. Wilson and Ensign Thompson, two sergeants and 34 rank and file; missing, Lieut. Wickham and three rank and file; 2nd Battalion, killed, two privates; wounded, Major Wood, Lieut. Saunders and 19 rank and file.*

For about ten days the army remained inactive, the inclemency of the weather and the difficulties of the country for transport necessitating a brief delay, after which a council of war decided to again resume offensive operations and attack

* W.O. 1, Book 180.

the Bergen position on October 2nd. Pulteney's column, consisting of three brigades of infantry including the 5th, 17th, and 35th Regiments, and two squadrons 18th Light Dragoons, with artillery, was in reserve, posted to cover the left of the English position, to threaten the enemy's right at Oudt Carspel, and to prevent him detaching troops to assist in the defence of Bergen. All through the night of the 1st-2nd of October, the army lay upon their arms, and about 2 a.m the columns silently moved off to their respective points of attack.

On the right, Abercromby's column captured Alkmaar, and eventually the French retired about 1 p.m., but although the British were victorious, the losses they had sustained (11 officers and 237 men killed, and 79 officers and 1,102 men wounded), added to the inadequate transport supply, the scarcity of water (the troops had none after 12 hours' marching and fighting, and the bad condition of the roads, caused by the continual heavy rains, delayed the arrival of supplies), prevented their success being followed up until the 6th, when a forward movement of the advanced posts was made, previous to an intended attack on the position of Beverwyck, which brought the enemy to oppose it, a general action ensuing, which ended in their being driven back to their original position.

In the action at Bergen on the 2nd October, the 2nd Battalion 17th had two rank and file killed, and Lieutenants Wynne and Morrison and five privates wounded.*

The situation of the army now became daily more critical and alarming.

Owing to the increased severity of the weather, it was with the greatest difficulty that the urgent necessaries of the troops could be procured, whilst the enemy in front held an almost impregnable position, fully confident of success, with a timely arrival of considerable reinforcements. Our army, moreover, was in a most exposed position on the unsheltered sandhills of North Holland, which no artificial means could improve, whilst, with no obstacle between them and the enemy, they were obliged to be constantly on the alert. Many of them had already been four days under arms in an almost incessant rain, and their remaining strength must soon be exhausted by

* W.O. 1, Book 180.

a continuance of such hardship and fatigue. So, considering all the circumstances, the Commander-in-Chief ordered a retirement to the old position of the Zuyper. Accordingly, on the evening of the 8th October, the right and centre of the army retired from their positions in front of Alkmaar, and reached the Zuyper by daylight, without the smallest molestation from the enemy. The left, under His Highness Prince William, having a longer march to make, retired gradually before the whole of the Dutch troops, and did not take up the posts allotted to them until the 11th and 12th.*

Shortly after this, the Dutch people not seconding these gallant efforts for their deliverance, negotiations were opened, which resulted in the allies agreeing to evacuate Holland by the 30th November, to restore 8,000 prisoners who were in England, as well as the guns and fortifications of the Helder, and to avoid opening the dykes to flood the country.

The books of the regiment were lost in this expedition.

To England Nov., 1799.

On the return of the army to England, the first embarkation of troops took place on the 22nd October, and the two battalions of the 17th were stationed at Dover. In a Return of Ships, which sailed with troops from the Texel on the 10th and 12th November, a part of the regiment is shown as sailing in the frigate "Venus," which put back on the 17th November;* both battalions had, however, reached Dover by the 23rd.

A War Office order, dated 7th December, directed both battalions to furnish parties to proceed to Dublin, of one field officer, two captains, four subalterns, six sergeants, 16 corporals, three drummers, and 24 steady men, to await orders from the Adjutant-General, and another order dated 9th December, directed the recruiting parties of both battalions to proceed to Ireland, consisting of one captain, three subalterns, with usual number of non-commissioned officers to each.

1800

Minorca had been captured in 1798 by General Sir Charles Stuart, the Commander-in-Chief in the Mediterranean, who, in December, 1799, submitted a scheme for a secret expedition to operate from it, in conjunction with the Austrians in Italy, against the French.

* W.O. 1, Book 182.

Minorca, 1800.

Both battalions of the regiment were selected to form part of this expedition, and, with the two battalions of the 40th, formed a brigade which embarked at Deal on the 27th and 28th March, and landed at Minorca on the 12th May. Both battalions of the 17th, on disembarking, were encamped at the entrance to the harbour, on the Lazaretto.*

A Monthly Return, dated, 1st June, shows the 1st Battalion 17th, with 443 effective rank and file, and 276 wanting to complete, and the 2nd Battalion 512, with 231 wanting to complete.

The 4th June being His Majesty's birthday, the usual celebrations took place, the two battalions 17th, with the 8th, 48th, and 90th Regiments, forming up at the east end of the town.

A War Office letter, dated, 11th June, directed that the 11th company of the 17th was to be discontinued on the establishment from the 25th May, inclusive.† Sir Ralph Abercromby arrived on the 22nd, and the whole of the troops, with the exception of the 8th, two battalions 17th, 50th, and 58th, left in Minorca as a garrison, were at once formed into brigades, and once more embarked on the fleet. With the exception of a few regiments left at Malta, this was the force with which Abercromby beat the French under Menou, near Alexandria (Egypt).

1801

This year both battalions of the 17th volunteered *en masse* for special service against the French in Egypt, as the following letters testify :—

"From—The Officer Commanding 1st Battalion 17th "Regiment, To—Lieut.-General the Honourable H. E. Fox, "Lieut-Governor of Minorca, dated, Ciudella, 25th April, 1801.

"It is with peculiar satisfaction I have the honour to "inform you that the first Battalion 17th Infantry possess "but one mind, in wishing to join their brother soldiers in "Egypt, and share their dangers and fatigues, and evidently "hope their services may be called forth for that purpose.

"(Signed) Richard Stovin, Lieut.-Col. Commdg.‡

* W.O. 1, Book 298.
† W.O. 4, Book 179, p. 230.
‡ W.O. 1, Book 299.

"Extract of a letter from Lieut.-General the Hon. H. E. Fox to the Rt. Hon. Henry Dundas (Secretary of State), dated Mahon, 25th April, 1801.

"It is with the greatest satisfaction I am to acquaint "you that the 1st and 2nd Battalions of the 17th Foot . . . "have in the handsomest manner, to a man, come forward "with an offer of their services for Egypt. I have thanked "them in the strongest terms for their spirit and ardour, "and accepted their services, in case they should now, or at "any future time, be called for.

"I have also assured them, that I should not lose a "moment's time through you and H.R.H. the Commander-in-"Chief, in laying before His Majesty their zeal and attachment "to his service. Indeed, too much praise cannot be given to "the non-commissioned officers and men on this occasion, "as I am confident it proceeded from themselves, for, "although the offer was made by the officers, they felt a "proper delicacy in not first suggesting it to them."

By Horse Guards letter, dated 22nd October, recruiting for the regiment was stopped, and the recruiting party directed to proceed to join the Army Depot at the Isle of Wight, for duty until further orders.

CHAPTER VIII.

Home Service and East Indies.

1802–1822.

1802

At the peace of Amiens, declared this year, England agreed to the restoration of Minorca, and the citadel of Port Mahon to Spain.

The Gazette of February 16th, 1802, contains a notice of prize money, due to the Army engaged in the expedition against Holland in 1799, under the command of the late Sir Ralph Abercromby: "The amounts are to be paid at the "White Horse Inn, Friday Street, London, commencing on "Thursday next, February 25th, and to continue until the "9th of March following."

State of Distribution.

General Officers, each	£231	4	3
Field Officers	62	8	3
Captains and others of that class ..	10	8	5
Lieutenants and Ensigns	4	14	3
Sergeants	2	0	7
Corporals, Drummers, and Privates ..		6	8

* * * * *

Monday, 8th March........1st and 2nd Battalion 17th Ft.

A General Order, dated, Horse Guards, 1st July, directed that an Armourer-Sergeant be added to the establishment of every regiment of cavalry and infantry.

reland, ıug., 1802

Both battalions of the regiment returned home, disembarking at the Cove of Cork, the 2nd Battalion on the 8th August,

OFFICERS' SHAKO PLATES (*exact sizes*).

1815 TO 12TH OCT. 1825.

1825 (13TH OCT.) TO 1828.

and the 1st Battalion on the following day, the regiment being taken on the strength of the Irish establishment from the 25th August.*

2nd Battalion disbanded.

A warrant from the Governor-General of Ireland, dated 17th August, directed that the 2nd Battalion was to be disbanded, and the officers placed on half-pay from the 25th September at the following daily rates, viz.: two lieut.-colonels at 8s. 6d.; one major, 7s. 6d.; six senior captains at 5s.; captain-lieut. and those junior and all lieuts. at 2s. 4d.; four ensigns at 1s. 10d.; paymaster, 7s. 6d.; adjutant and quarter-master, 2s. each; and surgeon, 5s.†

From Cork, the regiment (now a single battalion), marched to Limerick, and, on the 29th September, was inspected by Major-General Morrison.

A number of men having transferred their services from the 2nd Battalion, the parade state showed 1,014 effective rank and file, also 63 sick, 55 under size, and 27 unfit for service, or awaiting discharge.

On the 28th September the recruiting party of the regiment was ordered from the Isle of Wight to Ireland.

1803

A Royal Warrant, dated 30th May, directed that the regiment, from the 25th December, be placed on the establishment of 1,000 rank and file, with the usual proportion of non-commissioned officers and drummers.

From the earliest date of the formation of regiments in Great Britain, up to this period, it was ordained that colonels of regiments, and all field officers, should have troops and companies allotted to them, which, of course, in many instances they never did duty with.

A General Order, dated 6th June, stated :—

"His Majesty's pleasure, that in future each troop and "company throughout the army, shall have an effective captain, "and consequently, the field officers of the several regiments "of cavalry, foot guards, and infantry of the line shall no longer "have troops or companies."

* W.O. 8, Book 10, p. 103.
† W.O. 8, Book 10, pp. 53–55.

By the same General Order, the rank of Captain-Lieut. was abolished.

In July, the regiment was suddenly ordered from Limerick to Dublin, where a serious riot had taken place on the 23rd, when Lord Chief Justice Kilwarden and his nephew, the Rev. Richard Wolfe, were attacked in his carriage, and murdered by the rioters.

1804

England, April, 1804.

In April, the regiment marched from Dublin to Cork, and an Embarkation Return, dated 7th May, shows its embarking (28 officers and 902 men), in five transports for Newport, Isle of Wight, where it had a detachment at Sandown Barracks.

A War Office order, dated 18th June, directed that in consequence of the 17th proceeding to India, a recruiting company was to be added to the establishment from the 25th June, to consist of one captain, two lieutenants, one ensign, eight sergeants, eight corporals, and four drummers; total, 24.*

East Indies, July, 1804.

In July the regiment sailed for the East Indies, and arrived at Fort William on the 12th December, having lost Ensign Strickland by disease on the voyage.

1805

In June and July this year, the health of the officers and men suffered severely from the effects of the climate, all the officers except four, and upwards of 400 men, being at one period in hospital, though the mortality was by no means what might have been expected from such sickness. The casualties were replaced by a strong detachment from England, and the effective strength was augmented to 1,260 officers and men, which appears to have been the greatest strength the regiment ever attained during its service in India.

On the 12th September, the whole embarked in boats for the upper provinces; the headquarters and three companies landed, and encamped a short period at Allahabad, where they arrived on the 30th November, being intended as an escort to Sir George Barlow, who proposed making a tour

* W.O. 4, Book 193.

through that part of the country, but his presence being required at Calcutta, they proceeded to join the other companies which had marched to Cawnpore.

1806

Two companies were detached in October, 1806, with some artillery, under Captain Nicoll, to reduce a fort on the right bank of the Jumna; but it was evacuated by the insurgents on the approach of the detachment, and the two companies returned to their cantonments.

On the 20th of December, two companies of the regiment marched under Captain Hawkins, for the purpose of reducing several forts in the mountainous districts of Bundelkund, which was resigned to the British by the Mahrattas in 1804, but several chiefs proved refractory.

1807

Little resistance was made excepting at Chunar, which place was captured by storm in January, 1807, on which occasion Lieutenant Peter McGregor was killed, gallantly fighting in the breach. Lieutenant Despard also distinguished himself; he received a contusion in ascending the breach.

Captain Hawkins, referring in his report of the affair to Lieutenant McGregor, says: "Words are not adequate to "express the high sense all had of his bravery."

In an official communication on this subject, it was stated: "In justice to the merits of two companies of His "Majesty's 17th Regiment, forming part of the army imme- "diately employed in Coonch, Major-General Dowdesdell, "divesting his mind of every ground or intention of partiality, "cannot forego the expression of the sincere pleasure he has "received from numerous concurrent testimonies of their "excellent behaviour, which is generally allowed to have "been conspicuous in the affair at Chunar, and uniformly to "redound to the credit and highly appreciated character of "His Majesty's 17th Regiment."

The two companies returned to Cawnpore, where the regiment remained until September, when it commenced its march for Muttra.

In October, five companies took the field under Lieut.-Colonel Hardyman, for the purpose of attacking the fort of Comona, and on the 19th of November, the breach was stormed with great gallantry; but it was found impracticable, and the troops were forced to retire. The enemy was, however, so fully impressed with a sense of British valour and perseverance that he fled from the fort during the night. Captains Radcliff and Kirk, Lieutenants Harvey and Harrison, three sergeants, and 44 rank and file of the 17th were killed in the act of making a gallant effort to ascend the breach; at the same time Lieutenants Wilson, Campbell, and Dudingstone, two sergeants, 93 rank and file were wounded. Pay-sergeant Suttle distinguished himself, and was killed at the top of the breach. The conduct of the troops on this occasion was highly commended in General Orders by the Governor-General in Council; Lieut.-Colonel Hardyman, of the regiment, was particularly noticed.

The other companies of the 17th were afterwards withdrawn from Muttra, and the regiment was employed some time in pursuit of the hostile tribes, which made a short resistance at the fort of Gonoivie, and afterwards fled. The regiment then returned to Muttra.

1808

In November, 1808, the regiment joined the force under Major-General St. Leger advancing against the Sikhs, and proceeded as far as the River Sutlej, without meeting with serious opposition. The dispute being settled by negotiation, no action of importance took place.

1809

A War Office letter, dated 22nd March, to General Garth, authorised a reduction in the establishment of the regiment of 10 sergeants, 10 corporals, and 190 privates, from December 25th, 1808, the new establishment to consist of 1,008 rank and file.

In the beginning of May, the regiment returned to Muttra where it was increased by a large detachment from England,

and on 1st November, (800 strong), marched to Meerut, on Muttra being discontinued as a station for European troops.

The inspection took place at Meerut on the 18th December, by Major-General Fuller, when the good discipline of the regiment was commented on, and particularly its advance in line, and movements in double time.

1810

In June, the depot moved to Hinckley. This year the officers instituted a medal " For Military Merit," the ribbon of which was a warm shade of green, the medal here shown having been presented in the following year.

1811

The Inspection Report by Major-General Fuller, dated, at Meerut, on the 25th March, shows that some officers of the regiment still had outstanding claims for the Helder (1799) prize-money, which up to the above date had not been received.

On the 4th June, Colonel R. Stovin was promoted to the rank of Major-General, and was succeeded in the command by Colonel Hardyman, the second lieut.-colonel.

In September, the depot proceeded to Maldon.

1812

An Army circular dated 25th March, limited the award of corporal punishment, by regimental courts-martial, to 300 lashes.

At the inspection by Major-General Dyson Marshall on the 30th April, the marching in line of the regiment was reported on as " admirable."

The officers, this year, presented their late Colonel with a handsome and richly ornamented sword, which is described as " mounted with gold, beautifully decorated with military " devices and trophies, embossed and enamelled," with the following inscription, enamelled in gold on blade and hilt :—

" Presented to Major-General Stovin by the Officers of " the 17th Regiment of Foot, in which he served 33 years, as " a proof of their lasting esteem and regard."*

* R.U.S.I. Book of Cuttings, Vol. VI., p. 185.

On the 27th November, the regiment marched from Meerut to Ghazeepore, and in December, the depot moved from Maldon to Danbury.

1813

The regiment arrived at Ghazeepore in January, when two companies were detached to Chunar for fourteen months.

On the 23rd April, four companies, under Captain Despard, marched to Secrole and Mirzapore for the purpose of watching and intercepting the bands of marauders, called Pindarees, who infested the British territory, and these companies rejoined the regiment on the 18th June.

In consideration of the meritorious services of the non-commissioned officers of the army, and with a view to establishing in the infantry, a corresponding rank to that of Troop Sergeant-Major in the cavalry, a General Order, dated 6th July, directed that certain selected sergeants, with an increase of pay, should be called "Colour-Sergeants," the duty of attending the colours in the field being at all times performed by them, whilst it should in no way interfere with their other duties, the said sergeants to be distinguished by an honourable badge of a regimental colour supported by two crossed swords above a double gold chevron.

By an Order, dated 18th December, the depot, consisting of one field officer, one subaltern, and 40 men, was directed to proceed from Danbury to Portsmouth to join the Army Depot in the Isle of Wight.

1814

Large reductions of the infantry regiments took place this year. The majority of the second battalions, raised about the year 1782, were disbanded by War Office order, dated 12th October.

The depredations of the Nepaulese having brought on a war with that kingdom, the regiment was selected to join the army, invading in four divisions, that mountainous country. It commenced its march from Ghazeepore on the 31st October, and joined the division under Major-General J. S. Wood, which was ordered to take the field on the 15th November, and consisted of the following troops: His Majesty's 17th Regiment, a detachment of the 8th Native Cavalry, a due proportion of artillery, the left wings of both battalions 14th Native Infantry, 2nd Battalion 17th Native Infantry, and four companies each of the 8th and 12th Native Infantry.

Owing to the numerous obstacles to be overcome in the nature of the country, besides commissariat delays and the difficulty in getting coolies to carry supplies and baggage, the force did not arrive in the neighbourhood of Bhotwul (its objective point), until the 31st December, when scarcity of water caused a further delay in the advance. No enemy had as yet been seen, though it was reported they were moving down the Bhotwul Pass.

1815

Nothing of consequence occurred until the morning of the 3rd January, when the General decided on attacking the enemy, who were situated in the midst of a thick jungle, in the centre of which was a strong stockade. Our force was divided into two parties; one, consisting of six companies of His Majesty's 17th Regiment, nine companies of Sepoys with two guns, and some howitzers, was commanded by Colonel Hardyman (17th), and was accompanied by Major-General Wood; the other force, composed of two companies of His Majesty's 17th, and 300 Sepoys, was under the command of Major Comyn (17th Native Infantry), whose object was to turn the left of the Gurkha position, and thereby divert attention from the main attack.

The larger party marched about 9 a.m. on the 3rd, towards the position of the Gurkhas, situated about three miles from the English encampment, and the advanced guard was still in the thick of the forest, when a sudden turn in the road brought them in front of the stockade (an obstacle about eight feet high), and a heavy fire was at once opened on them which caused great confusion, through the stampede of frightened elephants in the jungle, whose mahouts had lost all control of them. The enemy followed this up with a sortie, which was repelled by the arrival of the main column, headed by His Majesty's 17th. The ground was rugged and the jungle thick, but our troops made a very gallant resistance, and kept up a warm fire on their opponents for over an hour. Meanwhile Captain W. Croker, who led the grenadiers, drove the enemy up the hill, killing a chief, Sooraj Tappah, and succeeded in ascending with his own, and one other company of the regiment, round the left flank of the enemy's work, the capture of which was apparently ensured, as the enemy were already retreating from it. General Wood, however, considering that the post, if carried, would be wholly untenable, determined to stop a fruitless waste of lives by sounding the "retreat." The whole of the troops engaged conducted themselves with great spirit, and could with difficulty be persuaded to withdraw from the enemy's stockade. When retiring, the light company of His Majesty's 17th made a gallant stand, enabling the remainder to withdraw in perfect order, and nearly the whole of this company were either killed or wounded, its commander, Lieutenant Poyntz, being shot through the chest.

The loss of the Nepaulese is said to have been little short of 500 killed and wounded. Several were shot on the tops of trees, whence they took aim at our men. Three chiefs were killed, one after having cut down two men of His Majesty's 17th. It is understood that our guides had proved treacherous, and had paid the penalty of their villainy with the loss of their lives.

The British casualties amounted to 24 killed, and five officers and 104 men wounded, of which His Majesty's 17th Regiment had 12 rank and file killed, and Lieutenants Pickering and Poyntz, and 57 rank and file wounded.

In a report on the action from Major-General Wood to the Adjutant-General (after referring to these officers having led the advanced parties of the column), he states :—

"Besides these officers with the light company, the "grenadiers and one battalion company, by an attempt to "turn the enemy's right, became exposed to a heavy and "galling fire. These companies, led by Captain Croker of "the grenadiers, partaking of the animated and spirited "conduct of their leader, and the other officers with them, "had nearly gained the crest of the works of Jutghur when, "finding that the enemy, from their incalculable numbers on

" the adjacent and commanding hills, would eventually over-
" power the detachment, I ordered the ' retreat ' to be sounded.

" The names of the officers of the companies alluded to
" are : Captain W. Croker, Lieutenants Young, Greenhill and
" Crawley of the grenadiers, Lieutenants Poyntz and Pickering
" of the light company, and Captain Halfhide of the battalion."

No further action took place with this division, and the rulers of Nepaul having been brought later to submission, the regiment, in May, returned to cantonments at Ghazeepore, and on the 28th October was inspected by Major-General Wood. The Inspecting Officer in his report remarked : " From
" my long acquaintance with this regiment, my personal
" knowledge of its gallantry in the field, and the uninterrupted
" harmony it enjoys in quarters, I am of opinion that His
" Majesty has few regiments that possess superior qualities."

The depot at Maldon was ordered in February to Portsmouth, *en route* to the Isle of Wight.

1816

At the inspection of the regiment, which took place at Ghazeepore, on the 30th April, by Major-General Wood, the following remark was made by him in the Inspection Report :
" With respect to every quality that is desirable in a soldier,
" whether in quarters or in the field, His Majesty's 17th Regi-
" ment is deserving of the highest commendation."

In 1816, a combination of native princes against British authority, called part of the regiment into the field ; in July of that year, the flank companies, under Captain Croker, proceeded to join a flank battalion forming at Allahabad, to unite with the army proceeding against Scindia, under the Marquis of Hastings.

It would appear that up to this period, regimental schools, in general, had not been conducted on any established principle, that of the regiment, during its stay at Ghazeepore, having been under the management of the army chaplain at that station who, (as the Inspection Report of the year says), " might have been
" considered its founder, on its present improved system."

On his being invalided this year, the supervision of it devolved on Lieutenant George Peevor of the regiment, who voluntarily took upon himself the instruction of the children and the young soldiers attending the school, to which he devoted all his spare time, with the result, that in each successive Inspection Report (from 1816 to date of the regiment returning home in 1823), this officer was specially brought to notice for his indefatigable exertions, in the benevolent cause he had undertaken. (See Plate 9.) In order to show the encouragement he gave to the pupils, the following regimental medal was instituted by him as a reward for merit.

Silver, 1¾ inches in diameter with blue ribbon.

Obverse: In the foreground, a tent with a palm tree behind it. In front is a female figure standing; her right hand holds a halberd, and her left is extended towards a soldier who is in the act of firing a field-piece; to the right is a stand of colours, crossed over a drum, guarded by a soldier shouldering a firelock. In the background, on a hill, is a fortification with a flag flying.

Reverse: Within a wreath, "The Reward of Merit." Legend, "H.M. 17th Regimental School, 1816."

The die of the obverse is preserved in the Mint at Calcutta. (See Plate 12.)*

On the 1st August the depot was ordered from the Isle of Wight to Chelmsford, and on the 12th November to Fort Pitt, Chatham.

1817

The battalion companies of the regiment, under Lieut.-Colonel Nicoll, formed, in October, 1817, part of the brigade under Brigadier-General Hardyman, which was ordered to proceed by forced marches towards Nagpore, where a body of British troops was surrounded. On the march, a considerable portion of the enemy's troops were discovered, in order of battle in front of Jubbulpore, with their right to the hills. The enemy's guns were captured by a charge of the 8th Native Cavalry, and the Arab infantry were attacked, overpowered, and driven from their ground with severe loss by the 17th Regiment. The two corps were thanked in General Orders for their distinguished conduct on this occasion. The 17th lost a few men and had Lieutenants Maw and Nicholson wounded.

* Tancred. p. 323.

The enemy evacuated the fortified town of Jubbulpore, leaving a quantity of stores, and the regiment continued its march towards Nagpore. Being obliged to halt two or three days at Lucknadoon, for the elephants to come up with provisions, information was received of the overthrow of the Nagpore Rajah's forces, and of the termination of his resistance; the regiment then returned to its cantonments at Ghazeepore; it received prize-money for the capture of Nagpore.

1818

The inspection took place at Ghazeepore, on the 8th May, by Brigadier-General Hardyman, when the regiment was reported on as "very efficient, and fit for any arduous service."

On the 25th December, it proceeded by water to Fort William.

1819

The regiment arrived at Fort William on the 24th January.

General Garth died, after commanding the regiment 26 years, and was succeeded by Lieut.-General Sir Josiah Champagné, G.C.H., from the 41st Regiment, on the 14th June.

In August, Colonel Hardyman was promoted to the rank of Major-General; he was universally esteemed as an officer and a gentleman, and the officers of the regiment resolved to present him with a sword, value £100, as a token of their respect; but proceeding to Meerut, to assume the command, he died suddenly of one of the diseases prevalent in that climate, before he received the sword, justly and universally lamented by all who knew him. He was succeeded by Lieut.-Colonel W. T. Edwards from the 73rd Regiment, the senior lieut.-colonel (Sir T. McMahon) being Adjutant-General of the Forces in India.

On the 10th August, the depot at Deal was ordered to Dover.

1820

On the 21st of December, the regiment marched for Burhampore, having lost, during the two years it was at

Calcutta, eight officers and 131 soldiers, cholera having been prevalent during that period.

In April, the depot was ordered to march from Dover, *en route* to the Albany Barracks, Isle of Wight.

1821

The regiment arrived at Burhampore on the 8th January.

On the 7th August two companies of the regiment, with five companies of Native Infantry and two 6-pounder guns, under command of Major Beck, 17th, attended the funeral, at Moorshedabad, of Shoojah-ool-Moolkh, late Soubah of Bengal. The detachment arrived early in the morning, and rendezvoused at the chowk* during the period of the ceremony of bathing the corpse. At 8, the procession started, preceded by the guns and the troops with reversed arms, the band of the regiment playing a solemn march, after which came the body, on a kind of bier covered with green velvet, on which was a canopy of the same material. Then followed the elephants, state horses, and troops of the deceased, with muffled kettledrums, the whole procession moving at a slow pace towards Jaffergunge, the burial place, about two miles distant from the city, and was attended, throughout the whole distance, by the Agent of the Governor-General, and the Civil Surgeon of the Station on foot. On arrival of the corpse at the place of interment, three volleys were fired by the troops, after which 29 minute guns, corresponding with the age of His Highness. The whole ceremony was conducted with faultless precision, through excellent arrangements made by Major Beck. The conduct of the troops was exemplary, and the regularity of their movements, after having performed a fatiguing march of nearly nine miles, was highly creditable to their officers.†

In November, the depot was attached to a Provisional Battalion of detachments at Portsmouth.

1822

By Horse Guards letter, dated 8th February, orders were received for the recruiting company of the regiment

* Bazaar. † MSS. Records, R.U.S.I.

to be discontinued on the establishment from the 25th January, its officers being placed on half-pay from the 25th March.

On the 10th August, the regiment proceeded by water to Calcutta, to embark for England.

On the 4th November, Colonel Edwards exchanged with Lieut.-Col. Archibald Maclean, of the 14th Regiment.

The regiment was inspected at Fort William on the 9th November, by Major-General Dalzell, and reported on as "a good and efficient battalion."

The following casualties took place during the Indian tour of service of the regiment, from 1804 to 1822 inclusive:—

	MEN
Lost by disease, and killed in action	1,021
Invalided	412
Transferred, or volunteered to other corps ..	452
Discharged on expiration of service	166
Lost by desertion	24

The total number of rank and file returning home was 309, and on the departure of the regiment, General Sir E. Paget, Commander-in-Chief in India, issued a General Order, in which he stated: "The Commander-in-Chief feels it to be "a just tribute to this old and distinguished corps to express "the high character it has always preserved in Europe, and "which, his Excellency is happy to find, has been maintained "during a long service of 18 years in India.

"A copy of this Order will be submitted to the gracious "notice of His Royal Highness the Duke of York, and the "Commander-in-Chief takes this opportunity of wishing the "regiment a prosperous voyage, and that it may long enjoy "its justly-earned reputation."

CHAPTER IX.

Home Service, New South Wales, East Indies, Afghanistan, East Indies, and Aden.

1823–1847.

1823

England, 1823.

On the 20th January, the regiment embarked at Fort William, sailed for England on the 30th, and after touching at St. Helena for a few days, landed at Gravesend on the 27th May, and marched to Chatham, after an absence of nineteen years from Europe, returning with four officers and 66 non-commissioned officers and privates of those who embarked with it in 1804.

Shortly after disembarking, orders were received for reducing the regiment to the home establishment, and on the 26th June, it was inspected by Major-General Lord Edward Somerset, when 259 men were recommended to be discharged, nearly all of whom were subsequently placed on the out-pension list of Chelsea Hospital.

The regiment, now reduced to 97 non-commissioned officers and men, marched to Portsmouth on the 6th July, and subsequently to Gosport, where it arrived on the 14th, and was joined by the depot, consisting of 205 non-commissioned officers and privates. On the 24th of October, it was reviewed on Southsea Common with the other troops at Portsmouth, Gosport, &c., by His Royal Highness the Duke of Clarence, afterwards King William IV., and, on the 12th November, marched in three divisions to Hull, where it arrived on the 3rd December, detaching one company to Carlisle and one to Tynemouth.

OFFICER'S SHAKO PLATE.
(*Exact size.*)
1829—45.

1824

This year the regiment commenced practising the new system of drill and field movements, as established in the army at this period, agreeably to the improvements introduced by Major-General Sir Henry Torrens, K.C.B., Adjutant-General of the Forces, and the sergeant-major and six non-commissioned officers were sent to London for instruction.

At the half-yearly inspection at Hull, on the 1st June, Major-General Sir John Bing, K.C.B., expressed himself much pleased at the proficiency of the corps in the new manœuvres, &c.

1825

On the 25th March, two companies were added to the establishment, which was increased to one colonel, one lieut.-colonel, two majors, 10 captains, 12 lieutenants, eight ensigns, five staff, 42 sergeants, 36 corporals, 14 drummers, and 704 privates.

On the 25th June, 1825, His Majesty King George IV. was graciously pleased to approve of the regiment " bearing " on its colours and appointments the figure of the ' Royal " ' Tiger,' with the word ' Hindoostan ' superscribed, as a " lasting testimony of the exemplary conduct of the corps " during the period of its service in India, from 1804 to " 1823."

Scotland, 1825.

From Hull the regiment was removed, on the 3rd July, to Scotland, the headquarters and six companies proceeding to Leith, and thence to Glasgow, and were transferred to Edinburgh for the winter, with a detachment at Stirling.

1826

On the 9th May, six companies, under Major Croker, were ordered to Paisley, and about this time the regiment unanimously subscribed one day's pay of all ranks towards the relief of distressed operatives of Paisley.

On the 17th July, orders were received for the regiment to march in three divisions to Greenock, where it embarked for Liverpool. It disembarked on the 29th, routes having been received for the march of detachments into Lancashire, where it was stationed three months, the headquarters being at Bolton.

Ireland, 1826.

On the 20th October, the regiment embarked at Liverpool for Ireland, and landed at Dublin, whence it marched in three divisions to Mullingar, with a subaltern's detachment at Granard.

From the following description of a medal, it seems clear that a regimental medal for ball-firing had this year been instituted.

Medal: *Obverse*, a tiger statant, supporting a shield inscribed "17"; below, "Hindoostan"; above, a crown; the whole surrounded by a laurel wreath. *Reverse*, "ball-" firing, 100 yards prize, Ensign D. Cooper, 1826." A circular silver engraved medal, $1\frac{7}{16}$ inches diameter, with a heavy raised floreated border, and chased ring for suspension.* Ensign D. Cooper, whose commission bore date 11th August, 1825, is shown in the Army List for 1826 as having received the Waterloo Medal.

1827

In April, the regiment was removed to Galway, where it is shown in the Monthly Return for May with detachments at 11 out-stations.

In October, the headquarters moved to Birr with a detachment at Roscrea, and in December to Templemore, supplying detachments to nine out-stations.

1828

On the 28th April, the regiment marched in three divisions to Dublin, and, by the 5th May, was concentrated in Richmond Barracks.

1829

England, 1829.

The half-yearly inspection at Dublin, by Major-General Dalbiac, which lasted three days, was concluded on the 9th May. The regiment was under arms from 8.30 a.m. to 12, when, shortly after, orders were received to embark at the shortest notice for England, the whole of the heavy baggage to be sent off at 4 p.m. and the regiment to march to the place of embark-

* D. Hastings Irwin, 4th Edition, p. 249.

ation at 6 p.m. To its great credit, this was effected without either drunkenness or absentees, and between 12 and 1 a.m. the two steamers on which it embarked sailed for Liverpool, arriving between 4 and 7 p.m. on the following evening (10th).

The headquarters and five companies went into billets at Liverpool, the left wing (five companies) disembarking and marching the same evening to Prescot, and, by the 24th May, the regiment was distributed as follows: headquarters and five companies at Rochdale, three companies at Halifax, and two at Bradford.

The Commander of the Forces in Ireland having expressed to the Commander-in-Chief his approbation of the order and regularity with which the embarkation was conducted, Lord Hill was pleased, in reply, to express the satisfaction with which he had received the report of the exemplary and excellent state of order and discipline in which the 17th Regiment embarked at Dublin for Liverpool, "a circumstance which he "considers highly creditable to the officers and soldiers of it."

On the 21st July, the headquarters marched from Rochdale *en route* to Chatham, where they arrived on the 7th August, when it was officially notified that the regiment was destined for New South Wales, as guards over convicts.

On the 30th July, Lieut.-Col. Maclean exchanged with Lieut.-Col. John Austin from half-pay, the latter retiring on the 13th August, when he was succeeded by Lieut.-Col. H. Despard.

The first guard, commanded by Lieut. Blackburne, embarked on the 12th November, and several other guards quickly followed.

1830

N.S. Wales, 1830.

On the 5th May, the regiment was inspected at Chatham by General Lord Hill, G.C.B., Commander-in-Chief.

On the 17th August, the headquarters, under Lieut.-Col. Despard, embarked on board the convict ship "York."

1831

The headquarters arrived at Sydney, New South Wales, on the 8th February, and proceeded at once to Paramatta

K

(where a detachment was left consisting of one field officer, one captain, one subaltern, and 35 men), and returned to Sydney on the 17th March.

Previous to the arrival of the headquarters in New South Wales, the regiment had detachments at Meriton Bay, Lower Portland Head, and Port Stephen, whilst one subaltern and 48 men were attached to do duty with the mounted police.

Most of the detachments that had been embarked as guards on board of ships, destined for Van Dieman's Land, were detained in that island and sent immediately on service into the interior, for the purpose of affording protection to the settlers, who were at the time suffering great depredations from the native islanders.

At one period, three captains, six subalterns, and upwards of 200 rank and file were detained on this service. The last guard, commanded by Major W. Croker, arrived at Van Dieman's Land on the 26th March, 1831.

1833

A Return of Military Officers, dated Sydney, 9th December, shows the following regimental officers in receipt of colonial allowances, with the appointments they held :—

Lieut.-Col. Despard, Commandant of the military district of Paramatta.

Major W. Croker, Commandant of Bathurst.

Captain Moffat, Police Magistrate at Port Stephen.

" Anley, Resident Magistrate at Maitland.

" Clunie, Commandant, Meriton Bay.

Lieut. Nagel, Superintendent Public Works, Meriton Bay.

Captain T. Williams, Commandant Mounted Police, and Lieuts. Darley and Steele, of the Mounted Police Division.

Lieut. Blackburne, previously employed on this duty was, on vacating it (3rd August), presented with a complimentary address by the inhabitants of the Hunter District.*

A General Order, published at Sydney on the 31st May, 1832, authorised a state of deductions to be made from the pay of officers holding colonial appointments of emolument, by Royal Warrant dated 31st December, 1830.†

* Newspaper cutting. † C.O. 201, Book 231.

1835

The regiment was reviewed by Major-General Sir R. Bourke, K.C.B., on the 21st May. In expressing in a General Order, dated Sydney, 22nd May, 1835, his entire satisfaction at the appearance of the regiment on the previous day, the inspecting officer further remarked: "The movements were "executed with great celerity and precision, and the march "in line merits particular commendation. The regimental "and company books are all well kept, and, in its interior "arrangements, this regiment has reached a high degree of "perfection."

1836

East Indies, 1836.

After occupying various stations in New South Wales for five years, the regiment received orders to proceed to India, and the headquarters embarked for Bombay on the 4th March; strength 7 officers, 3 staff, 14 sergeants, 7 drummers, and 240 rank and file.

On the 9th March, the second division of the regiment embarked, and two companies were left at Sydney until such time as their services could be dispensed with.

On the departure of the headquarter division from New South Wales, the General Officer Commanding, in a General Order dated Sydney, 3rd March, 1836, expressed his fullest approbation of the conduct of the regiment throughout its whole stay in the colony, and, in emphasising his opinion of its perfect state of efficiency for any service, he added: "In "whatever part of the globe His Majesty may require the "presence of the 17th Regiment, the Major-General feels "assured that it will maintain its credit and renown."

Lieut.-Col. Despard proceeded to England from New South Wales, on the 4th March, on retirement, by which the command devolved on Major W. Croker.

Previous to the regiment being placed on the Indian establishment from the 1st April, 1836, it had received the usual augmentation of one lieut.-colonel and 10 lieutenants.

The headquarters arrived at Bombay on the 14th May, and marched to Poona, where they arrived on the 27th, the second division of the regiment joining on the 14th June.

1837

The two remaining companies joined the headquarters at the camp near Poona on the 22nd January.

1838

The regiment was inspected at its encampment on the 17th January by Major-General Sir John Fitzgerald, K.C.B., when its condition was reported on as " highly creditable."

It remained at the camp near Poona until November, 1838, during which period events had transpired on the frontiers of Afghanistan, which, connected with the political measures of the chiefs who had assumed the dominion of that country, induced the British Government to undertake the restoration of the former sovereign, Shah Shooja-ool-Moolk, to the throne of that kingdom, as a precautionary measure, to protect the frontiers of the British dominions in the East against aggression.

Having been selected to join the force about to proceed on field service, the regiment marched to Bombay on the 8th November, mustering 30 officers, 33 sergeants, 10 drummers, and 552 rank and file, leaving at the depot at Poona three officers, six sergeants, three drummers, and 87 rank and file.

The following are the names of the officers who accompanied it :—

Lieut.-Col. W. Croker
Major Pennycuick
,, Deshon
Captain Darley
,, Miller
,, Erskine
,, Hackett
,, Bourchier
Lieutenant Owen, Acting Adjutant
,, Johnson, Brevet-Capt.
,, Dickson
,, Wetherall
,, Mathews
,, Clarke
,, De Teissier
Lieutenant Corry
,, Baird
,, Mauleverer
,, Bourke
,, Kyffin
,, Ruttledge
Ensign Welman
,, Cormick
,, Jones
,, J. L. Croker
,, E. Croker
Paymaster Moore
Quarter-Master Sarson
Surgeon Hamilton
Assistant-Surgeon Smith.

On the 20th November, the headquarters and first division of the regiment embarked for the mouth of the Indus, and the remainder a few days later. After a short passage they landed on the 2nd and 3rd December in Scinde, where the whole of the Bombay Division of the army assembled soon after under command of Lieut.-General Sir John Keane.

Brigadier-General Willshire commanded two brigades, and the 2nd Brigade was at first composed of H.M.'s 17th and the 19th and 23rd Regiments N.I., under Major-General Gordon, Indian Army.

On the 24th, the troops commenced their march northward, and advanced by brigades to the ancient town of Tatta, in Lower Scinde (situated on a rising ground four miles west of the river), where they arrived on the 28th.

1839

On the 23rd January, the march was resumed. Orders had been given for each man to carry a blanket, with a clean shirt, stockings, and flannel waistcoat wrapped in it, so that they might be enabled to change as soon as they arrived in camp after each day's march.

The weight of this, however, with the kit and 20 rounds of ammunition, the day's rations, and a small round keg, containing water, was no light burden for the men to carry in the heavy country through which they had come.*

Afghanistan, 1839.

To ensure the course of the Indus, the Bombay Division of the army, assembling for the invasion of Afghanistan, commenced its march from the mouth of that river, through the country occupied by the confederation of the Ameers of Scinde, who refused permission for the British troops to pass in peace through their territory, and a passage had to be effected by forcible means. Hyderabad, the capital, was captured.

In the diary of Colour-Sergeant John Clarke of the regiment, describing the march he writes: "We had to fight our "way to Hyderabad and halted to take the place, as the Rajah "would not allow us to march through his hunting grounds, "but when he saw our artillery pointed at his citadel he gave in, "When we left Hyderabad we had to march about 300 miles "through nothing but salt and sand."

*Col. Davis's History, "2nd Queen's," Vol. V., p. 7.

Kurrachee, the richest city of Scinde, was taken possession of, and the Ameers were brought to submission in the early part of February, 1839.

The army then continued its march, and the regiment reached Larkhana in Upper Scinde on the 4th March. It was from here that the sick of the European corps were sent up the Indus, by boats to Sukkar, under charge of Lieut. Corry of the 17th, who soon after lost his life, he and seven men of the regiment having died from heat and exposure on an expedition against a Beluchi fort, where upwards of 4,000 camels, stolen from us, and from the Bengal Force, were being harboured. A short time before, Lieut. Brady of the regiment, on his way to join head-quarters, had also died, with about 20 natives, from the effects of the sun.* The troops moved on from Larkhana on the 11th March, passed the great River Indus on a bridge of boats near the fortress of Bukkur, traversed an arid country to Usted, and afterwards marched through the desert plains of Baluchistan to Dadur, which they reached on the 5th April, occasionally suffering inconvenience from the want of water, and sustaining loss from the hordes of predatory natives.

From Dadur the troops marched through the Bolan Pass, with gloomy crags rising perpendicularly in awful grandeur on each side, to Dusht-i-be-doulut, or the Unhappy Desert, having some camp-followers murdered and baggage plundered, in these wild regions, by the Baluchis.

This occupied seven days, the pass being about 80 miles in length, and the road, for a considerable part of the way, nothing more than the stony channel of a mountain torrent, then partly dry.

On the 16th April, the regiment with its convoy marched to Quetta, daily fighting its way, and after continuing the march over difficult mountains and sterile plains, suffering from a deficiency of forage and provision, the army entered Afghanistan, when the Barukzye chiefs fled, and the British troops took possession of Candahar, the capital of Western Afghanistan.

Sir John Keane, having succeeded to the chief command of the army of the Indus whilst at Larkhana, the command

* Diary of Lieut. J. T. Mauleverer, who later, as Colonel, commanded the 30th Regiment.

of the Bombay column devolved on Brigadier-General Willshire. It was there also that the two Bombay brigades were broken up and the 1st Brigade now consisted of the Queen's and 17th only.

The regiment encamped on the grassy meadows of Candahar nearly two months, and the diary of Lieut. Mauleverer shows that in that interval the following were the exceptionally high prices of a few necessaries per pound, viz.: Coffee, Rs. 9; tea, Rs. 6; sugar, Rs. 4; flour and rice, each Rs. 1; grass sufficient for one horse for a day, Rs. 7½, and grain for ditto, 7 lbs. the rupee.

Leaving Candahar on the 30th June, the regiment, with the Bombay column, marched along a valley of dismal sterility to the Turnuck River; then advancing up the right bank, entered the country of the Ghilzees, and arrived before Ghuznee on the 21st July, where the whole of our army had assembled, Ghuznee being at this time a fortress of great strength, garrisoned by three thousand Afghans under Prince Mahomed Hyder Khan, well provided with stores, and every gate, excepting one, blocked up with masonry. Lieut. Mauleverer states: "The entire force marched at 4 a.m. on Ghuznee, and "we did not reach our ground until 1 p.m., having been "obliged to take a round through the hills of about ten miles, "to avoid the guns from the fort; 28 hours without a morsel "of food, and two nights under arms without tents."

SIEGE OF GHUZNEE.*

Arrangements having been made for storming the fortress, the troops selected for the operation silently took up the posts allotted to them an hour before daybreak on the 23rd July.

The storming column consisted of H.M's. 2nd, 13th and 17th Regiments, and the Bengal European Regiment. The latter, with the 2nd Queen's and 17th, was formed in column of sections on the road leading to the Cabul Gate, and within about 250 yards of it, preceded by the advance guard, consisting of the light companies of the three corps above mentioned, and a company of the 13th. The 13th Regiment was extended in advance of all, along the outside of the ditch of the fortress,

* The accompanying illustration shows a view of fortress from the spot where the Commander-in-Chief, Sir John Keane, and Staff witnessed the attack.

the artillery, in position on the left and right of the road, and the rest of the army in different parts in reserve.

The entrance was to be effected by blowing up the Cabul Gate with bags of gunpowder, and the engineer officer to whom this service was entrusted, succeeded in reaching the gate in safety, but some alarm being caused by the advance, blue lights were thrown up from the ramparts, and a heavy fire opened, which was promptly returned by our artillery. The fort was taken at 4.30 a.m. and upwards of 800* of the garrison killed.

The storming column in the meantime remained perfectly steady on the road, until a heavy explosion announced that the gate was blown open, when the whole rushed forward. The entrance was difficult owing to beams, stones, and other obstructions in the gateway. The advance guard, and in due course the storming column, succeeded in gaining a footing inside, after repelling one or two smart attacks by the enemy, sword in hand, and for a considerable time the struggle within the fort was desperate. In the frenzy of despair, the Afghans rushed out from their hiding places, and plied their sabres with terrible effect, but only to meet with fearful retribution from the musket fire or the British bayonets. There was horrible confusion and much carnage.

Some, in their frantic efforts to escape by the gateway, stumbled over the burning timbers, wounded and exhausted, and were slowly burnt to death ; some were bayoneted on the ground, whilst others were pursued and hunted into corners like mad dogs, and shot down with the curse and the prayer on their lips. Many an Afghan sold his life dearly, and though wounded and struck down, still cut out at the hated enemy.†

The 17th, having effected an entrance, formed in a small open space in the gateway, and from there, part of the regiment was detached to the right to clear some houses, from which a sharp fire was kept up, and part to the left, to clear the ramparts and streets in that direction.

The resistance met with was, in some instances, extremely obstinate. In one house alone, defended against No. 1 Company of the 17th, no less than 58 of the enemy were killed. Many other posts were defended with great resolution, but in less than an hour all resistance had ceased in the lower town.

* Lieut. Mauleverer. † Kaye's "Afghan War," Vol. I., p. 465.

The citadel, however, still held out, but on the approach of part of the 17th, under Lieut.-Colonel Croker, the gate was, after some parley, opened, and, a few minutes after, the colours of the regiment waved from the highest tower of Ghuznee, amidst the cheers of all ranks.

Lieut. Mauleverer describes the attack on Ghuznee as follows :—

"I do not think anything could have appeared more "magnificent to a passive spectator than the attack that night. "Just as the head of the column reached the gateway, at each "angle a tremendous blue light was burning, behind which, "stood out in strong relief, the gigantic outline of the fort, "with the lofty citadel frowning down on us. From every loop-"hole and embrasure, sheets of flame darted forth without "intermission, and lit up the dark and silent column below, "winding like a huge snake at the entrance to the dwelling "of the Prince. Behind us, our artillery fired ceaselessly, "every shot going over our heads. It was almost too grand "to comprehend at once; one was compelled to take it in detail."

As an instance of the spirit which animated the men, Dr. Kennedy mentions, that, on visiting the hospitals of Her Majesty's 2nd and 17th Regiments, he was surprised to find them cleared of sick. The gallant fellows had all but risen in mutiny against their surgeons, and insisted on joining their comrades. None remained in hospital but the hopelessly bed-ridden, who literally could not crawl; and even of these, a portion who could just stand and walk were dressed and made to look like soldiers, to take the hospital guard; no effective man could be kept away.*

Upwards of 500 of the garrison were buried by the besiegers, and many more were believed to have fallen beyond the walls under the sabres of British horsemen.

Sixteen hundred prisoners were taken, with immense stores of grain and flour, and a large number of horses and weapons.†

A standard was captured by the 17th which was afterwards lost in the wreck of the transport "Hannah."‡

The loss sustained by the regiment in the assault was comparatively small, being only one private killed and six

* C. Nash, p. 178. † Kaye's "Afghan War," Vol. 1, pp. 467—8.
‡ R. Cannon.

wounded, whilst the entire loss of the victors only amounted to 17 killed and 165 wounded, the latter including 18 officers.*

So doubtful was the success of the attack, and so very inadequate were our means, that Sir John Keane said at his own table that day, "Had we failed, nothing could have saved "the entire force from utter destruction."†

The following is an extract from General Orders issued by the Commander-in-Chief in India, on the occasion of the capture of Ghuznee, dated 23rd July, 1839.

"Lieut.-General Sir John Keane most heartily congratu-"lates the army he has the honour to command, on the signal "triumph they have this day obtained, in the capture of the "strong and important fortress of Ghuznee. His Excellency "feels that he can hardly do justice to the gallantry of the troops. "The advance guard under Lieut.-Colonel Denniss of H.M.'s "13th, consisting of the light companies of Her Majesty's 2nd "Queen's, 17th, and Bengal European Regiment, with one "company of the 13th, and the leading column, consisting of "Her Majesty's 2nd Queen's, 13th, 17th, and Bengal European "Regiments, &c. . . . . . To all these, and to other officers and "gallant soldiers, His Excellency's best thanks are tendered."

From Ghuznee the army advanced on the 31st July, on Cabul; the troops of Dost Mahomed Khan refused to fight in his cause, and the British, proceeding by triumphant marches to the capital, restored Shah Soojah-ool-Moolk to the capital of his dominions in the early part of August, the regiment arriving at Cabul on the 7th, with the Second Division, and encamped 6 miles from the city. Since entering the Bolan Pass half rations only had been issued, and for two months the men had not had a drop of grog, the greatest possible deprivation to a soldier when on service.

Lieut. Mauleverer's Diary here gives some instances of the adroitness of the camp thieves in the Cabul country as follows:—

"We halted last night (August 6th), five miles from the "Cabul city, and as thieves had been rather active of late, "one of our officers determined to use every precaution and "slept with a drawn sword in his bed, only to find, however, "that his camel trunks, containing his clothes, cooking pots,

* Kaye's "Afghan War," Vol. 1, pp. 467—8. † Diary of Lieut. Mauleverer.

" plates, silver spoons and forks, &c., had been stolen from his
" tent.

" The Adjutant-General of the Bengal Force, before we " joined them, was robbed, one dark night, of everything he " had in his tent. The thieves then took off his bed clothes, " tickled him on the soles of his feet, and when he jumped up " the rascals plucked the very sheet from under him and " bolted, leaving him regularly cleaned out.

" The Bombay Quarter-Master-General had ridden on to " Cabul with a strong escort, to examine the road, and took " a small tent to sleep in. When he awoke next morning, he " found that thieves had left him nothing but a red night-cap " and a pair of sleeping drawers."

Some days after the arrival of the 17th Regiment at Cabul, it was specially selected by the Commander-in-Chief, as a guard of honour, to receive Timar Khan, His Majesty Shah Soojah's son, on his arrival from Peshawur, and on this occasion the corps was highly complimented by Sir John Keane, on its clean, soldier-like, and efficient appearance.

On the 17th September, His Majesty Shah Soojah held a durbar in the Bala Hissar (the citadel of Cabul), when, amongst other officers, Lieut.-Colonel Croker, Majors Pennycuick and Deshon were nominated members of the order of the " Dooranee Empire," newly instituted by Shah Soojah, on being restored to the throne of Afghanistan.

This Order is divided into three classes, and was instituted by Shah Soojah-ool-Moolk, in gratitude to Great Britain for his restoration to his kingdom, and, also, as a reward to British officers by whom it was accomplished.

It is composed of gold, in the shape of a Maltese Cross resting on two crossed swords. In the centre, on a blue and green enamelled ground, are the words in Persian characters " Dur-i-Dauran " (Pearl of the Age), surrounded by a circle of pearls.* (See Plate 5).

On the 18th of September, the Bombay portion of the " Army of the Indus " left Cabul *en route* for India. The column reached Ghuznee by the same road it had advanced, and thence proceeded to Quetta, where it arrived on the 31st of October.

* Carter's "Medals of the British Army."

The regiment was afterwards detached, under Brigadier-General Willshire against Mehrab Khan, the Khan of Khelat, to reduce that treacherous chief to submission. It was plain that he was in no mood to submit to the terms dictated to him, and, doubting the intentions of the British to move against his stronghold, had been slow to adopt measures of defence. But when he knew that our troops were advancing on Khelat, he prepared like a brave man to meet his fate, and flung defiance at the infidel invaders.*

SIEGE OF KHELAT.†

The troops selected for this service consisted of H.M.'s 2nd and 17th Regiments, both very weak, and the 31st Bengal Infantry, in all about 1,000 bayonets with some light artillery and a few Irregular Horse. This force marched from Quetta, under Brigadier-General Willshire, on the 2nd November, 1839, and on the 12th reached Geranee, about eight miles from Khelat.

On the 13th, a little before sunrise, the column moved off, preceded by an advance guard, consisting of two companies of the 2nd Queen's and two of the 17th, the whole under command of Major Pennycuick (17th).

Having advanced about a mile, a body of the enemy's horse was observed at some distance on the right, moving nearly parallel with our line of march. Gradually approaching closer, they whirled suddenly to the left, advanced at a gallop and opened fire on the advance guard, when a skirmish commenced which lasted until the head of the column approached Khelat, when their skirmishers withdrew, taking the best shelter they could.

The Afghans were now observed in force, with five or six guns in position, on three heights outside the town, whilst the gardens on the left were occupied by them. The ramparts of the town and citadel were also manned, and every tower and housetop appeared crowded.

* Kaye's "Afghan War," Vol. II., p. 26.

† The view is taken from the hill at the north-west angle, which was occupied by the enemy; the Grenadiers of the 17th are advancing in skirmishing order. The headdress here shown of the rank and file is the old bell-topped shako of 1829, with a white cover over it.

The column halted about half a mile from their position, and arrangements were made for the attack.

The advance guard was immediately pushed forward to clear the gardens and enclosures on the left of the road, and this being done they took post in a garden about 300 yards from the walls of the town. The remainder of the troops, having been formed in three columns (the 17th in the centre column under Lieut.-Colonel Croker), advanced against the enemy's position under an annoying fire from their guns, until near the summit of the heights, when the three columns charged simultaneously, and each carried its height, the enemy retreating precipitately towards the town, leaving their guns behind them.

On perceiving this, the advance guard pushed rapidly across the intervening plain towards the gate, under a very heavy fire from the ramparts. The gate being closed before they could reach it, they were directed to take cover as best they could behind a low wall within 30 yards of it, and then to keep up a fire on the ramparts, in order to facilitate the approach of two guns advancing towards the gate with a view to blowing it open. Lieut.-Colonel Croker, with the remainder of the regiment under cover, was ready to push forward as soon as the chance of entrance should offer. The gate having been forced open by the fire of the guns, the assault is thus described by Brigadier-General Willshire in his despatch :—

" On observing this I rode down the hill towards the gate, " pointing to it, thereby announcing to the troops that it " was open. They instantly rose from their cover and rushed " in. Those under the command of Major Pennycuick being " the nearest, were the first to gain the entrance, headed by " the officer, the whole of the storming column from the three " regiments rapidly following and gaining an entrance as " quickly as possible under a heavy fire from the works, and " from the interior, the enemy making a most gallant and " determined resistance, disputing every inch of ground up to " the walls of the inner citadel."

The 17th, on entering, advanced along the main street, meeting with little opposition except at one place where a party of the enemy made an attack, sword in hand. They were, however, speedily disposed of, some being killed and the rest made prisoners. One of their leaders, after he had surrendered his sword to Colonel Croker (watching his opportunity, as the Colonel was turning away), suddenly seized him by the neck and sword arm, and threw him down, retaining his hold at the same time. The Colonel with difficulty extricated himself, and when he did he found his assailant despatched with several sword and bayonet wounds.

The regiment continued its progress through the town, under a galling fire from the housetops and by-streets, by which some loss was sustained, and Captain Bourchier, while clearing one of the streets with his company, was severely wounded. Prior to the regiment entering the town (by the north gate, on its being blown open), Major Deshon had been detached with two companies towards the south gate to cut off the enemy's retreat in that direction; Captain Darley, with the light company, being ordered to the right for the same purpose. In the meantime the advance guard, under Major Pennycuick (which had on first entering the town pushed on towards the citadel, repelling by the way one or two desperate attacks of the enemy, sword in hand), being joined by proportions of the other columns in its progress, succeeded in reaching the outer gate, which was forced open by applying the muzzles of about a dozen muskets to the lock or bar, and firing them off together.

Moving forward, the column entered a subterranean passage, perfectly dark and of considerable length, near the head of which a sudden rush was made by the enemy, and a scene of indescribable confusion followed. Order was, however, soon restored and the passage forced, the enemy retreating into a small court, near the head of it.

This court was occupied and most gallantly defended by Mehrab Khan in person, attended by most of his principal chiefs. A portion of the troops having discovered a passage above ground leading to the opposite side, a terrible slaughter ensued, Mehrab Khan himself and seven of his chiefs being killed with many of their followers. The upper part of the palace, overlooking the court in which this scene took place, was still defended, and the only access to it being by a narrow passage and staircase, completely commanded by the enemy's fire, some loss both in killed and wounded was sustained in repeated attempts to force it.

The Afghans, however, perceiving fresh preparations for another trial, expressed a wish to surrender on condition of having their lives saved. This, of course, was granted, and they accordingly about 4 p.m. delivered up their arms. They numbered about 80, and were commanded by Mahomed Houssein, the Wazeer, ranking next to the Khan.

The loss of the regiment, at the storming of Khelat, and in the previous affairs in the morning, was six privates killed, and one officer (Captain Bourchier), three sergeants and 29 rank and file wounded.

There was also, as at Ghuznee, a flag or standard taken by the regiment.

Referring to the amount of spoil which fell into our hands, Lieut. Mauleverer, in his Diary, mentions :—

" Immense quantities of flour, grain, provisions of all " sorts, bales of Indian silks, Cashmir shawls, Persian carpets, " English guns and pistols, Dollond's telescopes, china, glass, " and, in fact, a collection of many years, the fruits of spoil " and toll, levied on caravans going through the Bolan Pass, " were stored here, but very little money was found, although " there is supposed to be some."

The following is an extractfrom Column Orders, by Major-General Willshire, C.B., dated Camp, Khelat, 15th November, 1839 :—

" The preparation of the Major-General's despatches to " the Governor-General, reporting the brilliant achievements " of the force on the 13th instant, has prevented his attempt- " ing, until now, to express to the brave officers and soldiers " he has had the honour and happiness to command, his " unbounded admiration of their gallant conduct, and heartfelt " congratulations on the splendid triumph they obtained on the " 13th instant, by the dislodgment of the enemy from the " hills protecting the fortress of Khelat, and the immediate " capture of the place itself.

" As all was done in open day, and every individual had " the opportunity and gratification of seeing the uniform " steadiness and gallantry of all around him, it is only in the " power of the Major-General to tender to the brave officers " and soldiers engaged, and for them to accept, collectively " and individually, his warmest thanks and admiration of "their noble conduct, and to assure the officers that he "has not failed in his despatches to the Governor-General, to "mention the names of such as appeared to him proper to "do so."

Extract from Regimental Orders by Lieut.-Colonel Croker, dated Camp, Khelat, November 14th, 1839:—

"It is with the proud feelings of an old soldier, that "Lieut.-Col. Croker congratulates the gallant corps he has the "honour to command, upon the result of the attack upon "Khelat yesterday, when the cool, determined bravery and "admirable soldier-like conduct of all its members, who had "the good fortune to share in the operations of the day, were "so prominently conspicuous. To all, the Lieut.-Colonel "acknowledges the highest praise, and thanks are due for "their conduct on the occasion, but he would consider himself "wanting in justice did he not specifically mention Major "Pennycuick as having particularly distinguished himself at "the head of Nos. 6 and 8 Companies (part of the advance "guard of the army), which he so spiritedly led into the "town and citadel.

"Major Pennycuick having brought to the commanding "officer's notice the forward bravery of Colour-Sergeant J. "Dunn, and the Lieut.-Colonel having himself witnessed a "similar bearing on the part of Colour-Sergeant Mills "during the assault, he will perform a pleasing part of his "duty in entering their names in the records of the regiment "for the valuable example they respectively set to their "men.

"Lieut.-Colonel Croker cannot close this order without "an expression of deep regret for the loss the regiment has "sustained, in the deaths of six gallant soldiers who were killed, "and the sufferings of the equally gallant officer and soldiers "who were wounded."

The Chiefs, who had joined in hostile designs against the British interest, having been removed, and a friendly monarch placed on the throne of Afghanistan, a medal was given by the Government of India to the officers and soldiers present at the storming of Ghuznee, which the Queen authorised them to accept and wear.

Her Majesty Queen Victoria was graciously pleased to approve of the regiment bearing on its colours the words "AFGHANISTAN," "GHUZNEE," and "KHELAT," to commemorate its distinguished conduct in enduring the toils and privations of the campaign in Afghanistan with patient fortitude; its gallantry at the storming of Ghuznee on the 23rd of July; and its heroic conduct at the taking of Khelat on the 13th of November, 1839. Lieut.-Colonel Croker and Major Pennycuick were nominated Companions of the Order of the Bath; and the latter obtained the brevet rank of Lieut.-Colonel. Major Deshon received the brevet rank of Lieut.-Colonel and Captain Darley that of Major.

Soon after the capture of Khelat, the regiment continued its journey back to British territory in India, and arrived at Sukkur in Scinde on December 31st.

1840

General Sir Josiah Champagné, G.C.H.,* died on the 31st January, and was succeeded in the colonelcy of the regiment, on the 17th February, by General Sir Frederick Wetherall, G.C.H. (from the 62nd Regiment), an officer who commenced his military career in the 17th, and served with it during the American War.

East Indies, 1840.

On the 6th February, the regiment embarked in boats on the Indus and sailed to Tatta, where it arrived on the 13th; eight days later it marched to Kurrachee, and on the 16th March, the right wing and headquarters embarked on board the transport "Hannah," and next day sailed for Bombay.

THE WRECK OF THE "HANNAH."

The following extracts are from Lieut.-Colonel W. Croker's reports to the authorities.

Letter dated, "On board the 'Hannah,' 19th March, 1840.

"To the Deputy Adjutant-General, Queen's Troops,
"Bombay.

"I have the honour to report that I embarked with 300 "men of the regiment under my command, together with "50 native followers on Monday last. We sailed from

* Grand Cross of the Royal Hanoverian Guelphic Order.

"Kurrachee on Tuesday morning, 17th inst., and at about "9 that night ran on a sandbank under full sail, whilst we "were all on deck. Every endeavour has been made to get the "ship off, but without success. The whole of our camp "equipage, and a great part of our baggage, has been thrown "overboard, also the top-gallant masts, yards, sails, spars, "ballast, &c., and a quantity of the water started, to lighten "her as much as possible, but all to no purpose.

"Lieut. Jardine, Indian Navy, came on board this "morning, but was unable to persuade any of the country "boats to come off to our assistance, owing to the stormy "west wind blowing, and the heavy breakers amongst which "we are lying.

"I have, however, succeeded in sending off by the "ship's boats to the Kideywarry Mouth of the Indus, 95 "men, with a proportion of officers, under command of "Lieut.-Colonel Pennycuick, and am in hopes by this time "to-morrow to get off the remainder.

"I was forced to send the men away without their arms "and accoutrements, and shall, I fear, be obliged to sacri-"fice the whole of the latter, together with the new clothing "for the 300 men which is on board with us, my object "being to save the lives of all, as there is no saying how "long the ship may hold together, in the event of the wind "increasing."

In a second letter to the same address, dated, Kideywarry, mouth of the Indus, 21st March, Lieut.-Colonel Croker corroborates his former statement, as to the absolute necessity of leaving almost all the arms, &c., on board, and the total loss of the clothing. Then, after mentioning the receipt of some supplies, he continues :—

"The last of the troops, passengers and followers left "the wreck with me at 10 o'clock last night, at which time "the water was up to the main deck, and everything in the "hold afloat.

"All the troops are now assembled at the lower station "of the Indian Navy on this river, on a spot of land nearly "surrounded by water, and where we have just room to hut "ourselves, which we are doing as fast as the means to be "procured from the miserable villages around will admit of.

"The conduct of the troops during the trying period "from the night of the 17th to that of the 20th March, was "marked by the utmost order and regularity, the good effects "of which have been strongly exemplified by the fact that "not a single accident of any kind occurred.

"I cannot conclude without bringing to notice the "unremitting exertions of Lieut. Jardine, of the Indian Navy, "to get the troops on shore. He came off to the ship as "soon as he possibly could in a small boat, through a "heavy sea and frightful breakers, and from that period to "the present moment (for he has not yet been on shore), his "personal exertions in the boats were never wanting to urge "those under him to use their utmost endeavours to rescue us "from the dangerous situation we were in.

"Captain McGregor, the Commander of the 'Hannah,' "was on deck looking out when she struck, and I am happy to "be able to state that he did everything in his power to "get her off, and subsequently to save the people, which was "effected by his and Lieut. Jardine's boats."

On landing the men bivouacked on a sandy bank about four miles from the mouth of the river, the only dry spot to be found, the country for miles around being a perfect marsh. The officers experienced the hospitality of Lieut. Jardine, who had a tent pitched here. A few sheep and fowls for immediate use were procured from a village near, and on information of the shipwreck being received about twenty miles off, some native merchants were induced to bring in supplies of sorts.

The commissariat, too, at Tatta promptly despatched a boat with provisions for the troops.

Endeavours were now made to hire country boats for conveying the troops to Bombay, and, two boats having been procured, a detachment embarked on the night of the 25th. The night, however, proved so tempestuous that they were unable to get out to sea, and next morning, when preparing for another attempt, a steamer hove in sight, which proved to be the "Bernice," on her way to Kurrachee. Her commander, observing the wreck, stood in towards her, and, ascertaining what had occurred, took the whole of the officers and men on board, and proceeded to Bombay, where all arrived safely on the 31st March.

A report of the wreck of the "Hannah" in the "Bombay Gazette," March 30th, 1840, says :—

"The clothing of the 17th Regiment is all lost, and the arms "and accoutrements either lost or ruined. Treasure and mess "plate have shared the same fate. The Khelat jewels are gone, "and much prize property is said to be hopelessly lost, and "many valuables of the 'Heroes of Khelat,' worth thousands "of rupees."

The following is an extract from General Orders by the Governor in Council of Bombay, on the occasion of the wreck of the "Hannah."

"The Governor in Council has much pleasure in expressing "his approbation of the exemplary conduct of Her Majesty's "17th Regiment, under command of Lieut.-Colonel Croker, "on board the transport 'Hannah,' on the occasion of the "wreck of that vessel off Kurrachee, which, from the high state "of discipline it indicates, reflects the greatest credit on the "officers, non-commissioned officers, and soldiers of that regi-"ment. The promptitude and judgment of Commander Lowe, "of the Indian Navy, in returning to the Presidency with the "shipwrecked officers and troops, are approved by the "Honourable the Governor in Council, who also expresses "his approbation of the highly praiseworthy conduct of Lieut. "Jardine, who is reported, at considerable personal risk, to "have boarded the 'Hannah' when the ship was in the greatest "distress, and to have mainly contributed, by his exertions and "example, to rescue the troops from their perilous condition."

Extract from a letter to Lieut.-Colonel Croker, C.B., from the Deputy Assistant Adjutant-General, Her Majesty's Forces, on the same occasion :—

"Your own personal exertions, conduct, and good judg-"ment are fully appreciated by the Commander-in-Chief, and "the zeal of the officers has merited his fullest approbation, "whilst the orderly behaviour of the men under such trying "circumstances deserves every praise."

The left wing of the regiment, which had been left at Kurrachee, subsequently joined the headquarters at Bombay on the 12th April, and the regiment had only been assembled there eighteen days, when the headquarters and left wing proceeded to Poona, the right wing, under Lieut.-Colonel Pennycuick, remaining at Colaba. The latter had, however, joined headquarters by the 23rd May, as the report of the inspection (by Major-General Sir J. Fitzgerald, K.C.B.), which took place at camp near Poona, on that date, shows the strength on parade as 821 efficient rank and file.

1841

The regiment was inspected at camp near Poona, on the 23rd May, and reported on as "in perfect order, and well instructed in light infantry and outpost duties."

On the 12th June, the regiment marched to Bombay, where it arrived in ten days.

Aden, 1841.

On the 22nd September, the headquarters and four companies embarked at Bombay for Aden, where they arrived on the 2nd October.

From the time of the regiment being together at Bombay to the headquarters embarking for Aden, sickness had prevailed to a very great extent, and the casualties in consequence were numerous.

On the evening of the 5th October, 1841, a detachment amounting to about six hundred men, selected from the troops at Aden, proceeded, under the command of Lieut.-Colonel Pennycuick, to attack an Arab force which had caused much inconvenience by preventing supplies being received from the country. After a severe skirmish of two hours' duration, in the hottest part of the following day, the troops destroyed the Arab post of Sheik Othman, and returned to Aden on the evening of the 6th, having traversed upwards of forty miles of ground in about twenty-two hours.

On this occasion an artillery officer and four privates were wounded, and one died from sunstroke. The Arabs were reported to have lost 17 killed, and the number of their wounded was not ascertained.

An extract from a letter to the Political Agent at Aden, from the Secretary to the Government of Bombay, states:—

"The expedition seems to have been well conducted, "and Lieut.-Colonel Pennycuick and the officers and men under "his command are entitled to the approbation of the Government for their zealous exertions."

1842

The headquarters of the regiment remained at Aden during the year 1842. In February, a detachment proceeded from Bombay to Poona, and in November, a detachment marched from Poona to Ahmednuggur.

On the 18th of December, 1842, the venerable General Sir Frederick Augustus Wetherall, G.C.H., died, after a service of sixty-seven years, and attaining the age of eighty-eight.

The number of deaths in the battalion from the 31st March, 1840, to the end of 1842, amounted to four officers and 305 rank and file.

1843

Her Majesty Queen Victoria was pleased to confer the colonelcy of the regiment on Lieut.-General Sir Peregrine Maitland, K.C.B., from the 76th Regiment, on the 2nd of January.

1844

The headquarters of the 17th remained at Aden throughout the year.

On the 5th July, Viscount Hardinge (on his way to India, as Governor-General), disembarked at Aden and inspected the battalion, reporting very favourably on the splendid physique of the men.*

In July and August, 1844, the detached wing at Ahmednuggur was affected with cholera. In the course of fifteen days one hundred and eight cases occurred; the deaths during the period amounted to thirty-two. Amongst them was Brevet Lieut.-Colonel Deshon, an officer of the highest talents and character, and public and private worth.

During the latter part of the year 1844, and beginning of 1845, a company of the regiment was employed on field service in the southern Mahratta country, in the operations before the forts of Monohur and Munsuntosh, where, at the assault of the latter, four privates were killed; Lieut. Gardiner, who commanded the company, one sergeant, and five rank and file were wounded.

* "Life of Viscount Hardinge," p. 57. By his Son.

The officer commanding the force, in a field detachment order, expressed his high sense of the gallant bearing of the officers and men of this company, on every occasion of their meeting the enemy, and of the cheerfulness shown under the privations of fatigue, to which they were frequently exposed in the harassing service they had to perform during the four months they were under his command, and he brought to notice in a despatch, the valuable services rendered by Lieutenants Gardiner and Belfield, Ensign McCrea, Assistant-Surgeon Willis, and the non-commissioned officers and men of No. 8 Company, 17th Regiment, all of whom in due course received, in a despatch, the approbation of the Government of India, and of the Commander-in-Chief, for the conduct they displayed.

1845

The authorised establishment of the battalion in January, this year, was 10 companies, and a total of 1000 rank and file.

East Indies, 1845.

The headquarters of the regiment embarked at Aden on the 13th of March, 1845, and arrived at Bombay on the 6th of April. The left wing marched from Ahmednuggur on the 11th of December, and joined the headquarters at Bombay on the 26th of the same month.

1846

On the 2nd January, the regiment, having been selected for field service, embarked at Bombay for Scinde, and on the 11th of January marched from Kurrachee, *en route* to Bhawulpore; it arrived at Sukkur on the 3rd of February, and on the 16th of that month proceeded on its march towards the Punjaub; but accounts being received of the termination of the war in that country, the troops advanced no further than Bhawulpore, on the Sutlej, where the regiment remained until the 12th March, when it returned to Sukkur, and was in camp there until the 31st, when it moved into barracks.

The regiment was inspected at Sukkur on the 21st April, by Major-General Hunter, whose report remarked on their "exemplary and soldier-like conduct, both in the field and "quarters."

From the middle of May to the end of July, the corps suffered much from apoplectic fever, the casualties during this period amounting to forty-eight.

The headquarters embarked for Kurrachee on the 9th August, landing on the 14th, and, after being encamped for a month, moved into barracks.

The following statement shows the effective strength of the regiment, exclusive of officers, on its arrival in India, with the accessions and casualties, which took place during the period of its service in that country, viz., from the 14th May, 1836, to the 13th March, 1847.

Strength on arrival in India, 14th May, 1836	631	Died in India, Afghanistan, Baluchistan, Scinde, &c., including killed in action ..	680
Recruits joined from that period until the embarkation of the regiment for England, 13th March, 1847	1,222	Discharged and invalided ..	376
		Transported	4
		Missing on march to Cabul ..	1
		Promoted to a commission ..	1
Volunteers and transfers received from other corps ..	130	Volunteered and transferred to other corps	447
		Embarked for England with the regiment, including detachment, on board the "Malabar"	474
Total	1,983	Total	1,983

Of the officers and men who embarked with the regiment on service, in November, 1838, 20 officers and 472 men returned with it to Bombay in March and April, 1840. The loss by death, including killed in action, during the Afghan campaign, was 92, including three officers.

On the 10th October, a General Order was issued, permitting men wishing to remain in India to volunteer to other corps prior to the regiment embarking for England, when upwards of 233 men volunteered, chiefly for the 8th, 28th, and 78th Regiments.

A further opportunity for volunteering having been given in January, 217 more took advantage of it, leaving the effective strength of the regiment, in sergeants, drummers, and rank and file, a total of 474, including a small detachment previously sent to England on recruiting service.

The following is a copy of a General Order, issued by Lieut.-General Sir Thomas McMahon, on the regiment leaving India, dated Bombay, 8th March, 1847:—

"The Honourable the Governor in Council, having, with "the sanction of the Government of India, directed the Com-"mander-in-Chief to make the necessary arrangements for the "embarkation of Her Majesty's 17th Regiment for England, His "Excellency has much satisfaction in availing himself of the "opportunity to notify his unqualified admiration and respect "for this distinguished corps, which is now about to return "home, after an absence of seventeen years, eleven of which "have been passed in this Presidency, being the second tour of "service in India which the regiment has performed within the "half-century.

"The public records bear ample testimony to the soldier-"like conduct in quarters and conspicuous gallantry in the field "which the regiment has always evinced, and, in the unim-"paired good order which characterises the corps at the present "moment, it is peculiarly gratifying to His Excellency to "observe, how successfully a high state of discipline may be "maintained in a regiment throughout many years of varied "service in India.

"This creditable condition can only have been attained by "the exertions and vigilant superintendence of the commanding "officer, combined with the zealous support of the officers and "non-commissioned officers, and in cordially acknowledging "the claims to approbation which all ranks have established, "His Excellency takes this occasion to mention, as specially "deserving thereof, their old and meritorious commanding "officer, Colonel Croker, who has passed the whole of his "long and honourable military career in this one corps, and "who, by his good judgment and ability, no less than by his "gallantry and the energetic discharge of his duty in all "respects, has mainly contributed to raise the regiment to "the high character which it bears in the service.

"His Excellency also deems it but just to name Lieut.-"Colonel Pennycuick, the second lieut.-colonel of the 17th "Foot, an officer of tried capacity, as particularly deserving of "commendation, on the various occasions when the command "of the regiment has devolved upon him.

"But ably as it has always been commanded, the admirable "efficiency to which it has arrived reflects also, in His Excel- "lency's opinion, the highest credit upon the corps at large, and "as, after a long acquaintance with the regiment, his intercourse "with it is now about to cease, its merits alone, independent "of other considerations, cause His Excellency to part with it, "with feelings of the warmest interest in its future welfare, "and with the fullest confidence that success and glory will "attend it wherever its services may be required."

OFFICERS' BREASTPLATES.

(*Exact sizes.*)

1825

1831

CHAPTER X.

HOME SERVICE, MEDITERRANEAN, AND CRIMEA.

1847–1856.

1847

England, 1847.

THE regiment embarked at Bombay on the 13th March, under command of Lieut.-Colonel Pennycuick, K.H., in the freight-ships "Ann" and "John Brewer," and next day sailed for England, arriving at Gravesend on the 6th August, and disembarking on the 7th, marched to Chatham and thence to Canterbury, where it was concentrated by the 13th.

On the 10th November, the regiment was moved to Dover.

On the 5th November, Colonel Croker retired by the sale of his commission in favour of Lieut.-Colonel Pennycuick.

1848

On the 7th April, Lieut.-Colonel Pennycuick exchanged to the 24th Regiment with Lieut.-Colonel Stoyte.

Apprehensions were entertained that the public peace would be disturbed by the several meetings of Chartists in the vicinity of the Metropolis on Monday, the 10th of April, 1848; and as they appeared determined to unite on Kennington Common, in order to proceed thence, in procession, to the House of Commons with their petition, the Government took the usual precautionary measures to prevent tumultuous assemblages of the people. Accordingly the regiment was ordered to proceed from Dover to London on the 8th of April, but happily the meetings dispersed more quietly than was anticipated, and the regiment marched to Portsmouth on the 13th of that month, in which garrison it remained until the 26th of July, when it proceeded to Chatham.

1849

On the 11th April, the regiment marched in two divisions to Canterbury, and returned on the 21st and 22nd June to Chatham. Whilst stationed at these places it furnished a one-company detachment to Harwich and three companies to Sheerness.

1850

reland, 1850.

On the 3rd, 4th, and 5th April, the regiment moved to Weedon with detachments at Wolverhampton and Northampton until the 6th June, when it embarked at Liverpool for Ireland, the headquarters marching to Castlebar with detachments at four out-stations, and thence to Galway, when a further re-distribution of detachments took place.

1851

The regiment was inspected at Galway by Major-General Napier, C.B., on the 27th May and 22nd October, and reported on as " in excellent order and drill particularly good."

1852

On the 26th March, the headquarters and three companies proceeded by rail to Dublin, and occupied quarters in Richmond Barracks, the detachments joining soon after.

1853

On the 24th March, the regiment moved to the Palatine Square, Royal Barracks, under command of Lieut.-Colonel M'Pherson, C.B., Colonel Stoyte having been appointed Inspecting Field Officer of the York Recruiting District.

At the half-yearly inspection in Dublin on the 10th October, Major-General Cochrane reported the regiment as " in a very efficient state, and upon the whole I do not believe " there is a regiment in the service in better condition."

On the 13th October, orders were received for the regiment to proceed to the Mediterranean.

1854

On the 16th February, the regiment was moved to Templemore until further orders.

On the 28th April, the headquarters and three companies, having embarked at Cork, sailed for Gibraltar under command of Lieut.-Colonel P. M'Pherson, C.B., the left wing following under Major Bourke, the whole regiment having arrived there by the 13th May.

Gibraltar April, 1854.

At the half-yearly inspection on the 18th May, by Lieut.-General Sir Robert Gardiner, K.C.B., he expressed great satisfaction at the appearance of the regiment, both in the field and in quarters.

On the 31st May, Lieut.-General T. J. Wemyss, C.B., was appointed Colonel of the regiment, in place of Lieut.-General Sir Peregrine Maitland, K.C.B., deceased on the previous day.

By War Office Order of the 13th November, the establishment was augmented to six depot and six service companies, with a total (officers and men) of 1,200, which later in the month was increased to eight service and four depot companies, with a total of 1,400.

On the 27th November, the regiment received orders to reinforce the army serving under Field-Marshal Lord Raglan in the east.

The following is an extract from the garrison orders, issued at Gibraltar on the occasion of the departure of the regiment for the east, dated 30th November, 1854 :—

" The Governor, under other circumstances, would have
" greatly regretted the departure of those regiments from the
" garrison. During the period of their serving together, their
" conduct has been marked by exemplary order and regularity,
" and calls for that unqualified praise with which it will be his
" duty to speak of them, in reporting their departure to the
" Commander-in-Chief as regiments of such marked discipline
" in garrison.

" The Governor confidently foretells their distinction in
" the glorious field they are called to. That field has already
" been marked by victories which will ever stand prominent
" among England's proudest triumphs, but yet much remains
" to be done, calling for endurance under chance privations,

"and those qualities of energy and ardour with cool determina-"tion which characterise the British soldier. They will enter "upon the glorious field before them with every incentive "that animates soldiers: their Sovereign's prayer and "benediction, their country's pride and fervid hope, the good-"will and eyes of all Europe fixed on them, as engaged in the "sacred defence of a righteous cause, against an unprincipled "and aggressive ambition, aiming at its ends through a "barbarous and ruthless war.

"The Governor takes leave of his brother-officers and "soldiers with the truest attachment and ardent wishes for "their health, and his assurances that he will never cease to "feel the deepest interest in whatever can exalt their names "in the service, or promote their individual welfare and "happiness."

mbark for rimea.

On the 2nd December, the regiment (eight companies, consisting of 23 officers, 41 sergeants, 38 corporals, 17 drummers and 621 privates) embarked under Colonel P. M'Pherson, C.B., in the steamship "Tamar" for the Crimea, reaching Balaklava Harbour on the 16th December.

The following are the names of the officers who arrived there with it:—

Colonel P. M'Pherson, C.B.

Major O. P. Bourke	Lieut. J. B. H. Boyd
Captain T. O. Ruttledge	,, F. W. Lukin
,, J. L. Croker	,, R. Swire
,, E. Croker	,, C. M'Pherson
,, W. Gordon	,, A. P. Traherne
,, A. M'Kinstry	,, E. F. McBayne
,, R. J. O'Conor	Ensign C. G. Grant
Lieut. W. H. Earle	Paymaster W. Telford
,, T. H. Brinckman	Lieut. and Adjt. C. H. J. Heigham
,, R. E. Williams	Quartermaster J. Campbell
,, R. Smyth	Surgeon W. Simpson

The regiment disembarked on the 17th December, and after encamping that night on the spur of the hills, at the head of the harbour, marched, *en route* to the front, to Karani (four miles), and encamped for the night. The route was most difficult and fatiguing; the men, besides their ordinary field equipment, carried two days' cooked provisions and the whole of the camp equipage, such as tents, &c., whilst the steepness of the way, over a range of hills saturated with rain, and of a most tenacious and slippery mud, rendered their progress almost impossible. Cholera broke out, from which five men died. Next day (19th) the regiment marched to the camp of the 4th Division, and were attached to the 1st Brigade of it.

Here Colonel P. M'Pherson took command of the 1st Brigade, handing over the regiment to Major Bourke.

On Christmas night, 1854, the regiment furnished for the first time, a party for the Greenhill Trenches (Left Attack), consisting of four officers and 150 men.

1855

On the 6th January, the establishment of the regiment was increased to eight service and eight depot companies, with a total (officers and men) of 2,000, and on the 19th March, this latter strength was distributed as follows: In the Crimea eight companies, at Malta four, and at home four.

By the 26th January, a draft had arrived at Balaklava in two ships consisting of 281 rank and file and the following officers: Major Cole, Captains Colthurst and Brice, Lieutenants Dyer, Tompson, Disbrowe, Lees, Butler, and J. O. Travers, and encamped on a spur of the hills at the head of the harbour. The situation was very exposed and the cold intense, the thermometer one night falling to six degrees below zero. Here the men were provided with warm clothing, and, as far as possible, long boots.

Major Cole immediately proceeded to the front and assumed command of the regiment, relieving Major Bourke of his temporary command.

On the 12th February, the regiment marched to Balaklava, and exchanged the common musket for the new pattern Enfield rifle of 1853.

On the 22nd March, a general sortie upon the whole of the besiegers' entrenchments was made by night, which was gallantly repulsed at all points; two captains (E. Croker and Heigham), three subalterns, and 398 rank and file of the regiment took part in it.

On the night of the 12th April, a special fatigue party, consisting of four officers, eight sergeants, and 200 rank and file, was sent, with half that number of the 57th Regiment, into the Greenhill Trenches, for the purpose of dragging guns into a new battery in the advanced trenches (a similar party of other corps having failed the night before in effecting this in consequence of the deep mud and the heavy fire of the enemy). This they effected in spite of these difficulties, and Colonel M'Pherson, commanding the brigade, was directed by Lord Raglan to thank the officers and men thus employed publicly, in his name, for their gallantry and perseverance on this occasion.

On the 16th June, Colonel M'Pherson having been invalided to England, Major-General Sir John Campbell, Bart., was appointed to command the 1st Brigade of the 4th Division.

Assault on the Redan.

At midnight, on the 17th June, the above brigade, under Major-General Sir J. Campbell, paraded for the assault on the Great Redan.

The 17th Regiment, having been in the trenches (Left Attack) the whole of the previous day, only returned to camp at 9.30 p.m., and after a meal and an hour's sleep marched down to the quarries on the Right Attack, nearly 800 strong of all ranks, under Lieut.-Colonel Cole. The 57th Regiment led, with a party of riflemen from the 2nd Battalion Rifle Brigade, followed by the 17th, 21st Fusiliers, and a working party of 400 rank and file of the 20th Regiment.

At daylight the brigade was in its appointed place, but owing to a misconception of the orders, want of knowledge of the locality, no proper guide, and the utter darkness of the night, the column went into the advanced trenches in front of the quarries, instead of the open space in rear of them. In consequence of this error the troops could only advance in file, instead of in quarter-column as directed. The enemy was, as it appeared, aware of the intention of the besiegers, and, the moment the head of the column showed itself outside the trenches, a terrific fire of shot and shell and musketry was poured upon them from the front and both flanks.

The Major-General and many other officers, with a large number of men, fell; and this column, although many had reached the abattis, was driven back to the trenches. As soon as Sir John Campbell's death was announced, Lieut.-Colonel Lord West, being the senior officer present, determined to make another attempt. The grenadier company of the 17th, having been formed in the circular advanced trench, rushed over the parapet, whilst the remaining companies, with sailors carrying ladders, and the 57th, passed out by an opening on the left of that trench, but the fire (which, heavy as it had been before, had now increased into a perfect hurricane of iron) withered the head of the column.

As the trenches were narrow and intricate, and moreover choked with the dead and dying of the previous attempt, the men could not be brought out in sufficient numbers, whilst the heavy cross fire effectually prevented anything like a formation. A second time the column was driven back, and, some hours afterwards, orders were sent for the 1st Brigade to be relieved by the Light Division, and the men returned thoroughly exhausted by nearly thirty hours of hard work under fire. The loss in the brigade was severe, but the heaviest fell on the 57th Regiment. The 17th lost Captain John Croker, one sergeant, and 12 rank and file killed; five sergeants and 27 rank and file wounded.

Lieut.-Colonel Cole, Captains Gordon, M'Kinstry, and O'Conor, and Lieutenant Tompson were favourably mentioned in despatches, the first being nominated a Companion of the Bath, and the second to a brevet majority for their services. The name of Lieutenant J. O. Travers had been accidentally omitted in the despatch, his gallantry having been remarked by all who saw him.

The following Regimental Order was issued by Lieut.-Colonel Cole on this occasion, dated, Camp before Sevastopol, 18th June, 1855:—

" Lieut.-Colonel Cole cannot pass over the present occasion
" without expressing his deep gratification at the conduct of
" the regiment on the late trying occasion, when the whole
" of the regiment, both officers and men, behaved with a
" steadiness and courage which could not be but highly
" pleasing to him, and he cannot but express his grateful

"thanks to them, one and all. The grenadier company "especially behaved with a gallantry which he owns he "expected from them, and they, in common with the rest of "the regiment, will join him in deeply deploring the loss "sustained in Captain J. L. Croker, whose conduct was beyond "praise.

"He desires to compliment Lieutenants Tompson and "Travers on their conduct on this trying occasion, also "Sergeant John Plant and Privates Bourke, Dowdle, Eares, "Lawless, and Thompson, who most gallantly recovered the "body of their lamented captain under the heaviest fire. "These officers and men may rely that he will not forget "their conduct."

Extract from General Orders, dated, Headquarters, Sevastopol, 28th June, 1855: "The Field-Marshal "has the satisfaction of publishing to the Army the "following extract from a telegraphic despatch, dated 2nd "June:—

"'I have Her Majesty's commands to express her grief "'that so much bravery should not have been rewarded with "'merited success, and to assure her troops that Her Majesty's "'confidence in them is entire.'"

The undermentioned men of the regiment also distinguished themselves at the assault of the Great Redan on the 18th June, 1855:—

Private Thomas Code „ Chas. E. Booth	By going out over the parapet, after the repulse, and bringing in, with others, the body of Captain Croker.
Private Dunbar Grisby -	By going out and bringing in the remains of Sergeant Connell.

Corporal Philip Smith, who went over the parapet several times and brought wounded men on his back into the trenches.

Lieut.-Colonel Cole was invalided to England by a Medical Board on the 18th June, and Major Ruttledge assumed command of the battalion, until himself invalided on the 23rd, when the command devolved on Captain E. Croker until the 30th July.

Up to the 29th June, Captain Armstrong, Lieutenants Versturme, Macreight, and Dyer, two sergeants, two corporals, and 98 privates had joined the service companies from England and from the reserve companies at Malta, and on the 11th July, Lieutenant A. H. Utterson joined.

On the 6th September, a draft joined the regiment consisting of Captains P. M'Pherson and Lindesay, Lieutenants Robinson and Parker, four sergeants, seven corporals, and 91 privates, followed, on the 12th, by Ensigns Webber and Hartwell.

On the 8th September, the regiment, under command of Brevet-Major W. Gordon, was engaged in the final assault of the Great Redan.

The 4th Division (in which the 17th was the senior regiment), under the command of Major-General Bentinck, was detailed to act as a reserve to the attacking column, but, soon after the attack commenced, the regiment was ordered to advance, which it did under an exceedingly heavy fire, and had reached the front trenches, and was forming to enter and support the troops in the Redan, when the " Retreat " was sounded and the attack given up for a time.

The regiment, with the rest of the 1st Brigade, was ordered to remain in and guard the trenches during the night, and so had the gratification of witnessing the destruction of the enemy's magazines, &c., previous to their evacuating the town that night, after which a party of the regiment was detailed to bring in the wounded officers and soldiers from the Redan.

At 7 a.m. the following morning the regiment was relieved, and marched to the camp.

The casualties during the day were : Lieutenant Tompson dangerously, and Lieutenant Swire severely wounded, Lieutenant Parker slightly wounded, one man killed, and 19 wounded. Major Gordon and Lieutenant Swire were favourably mentioned in despatches.

In commemoration of the fall of Sevastopol, and the part the regiment took in its protracted and harassing siege, the word " Sevastopol " was granted to be borne on its colours.

The following telegraphic message from the Secretary at War was published to the troops by the Commander of the Forces :—

" The Queen has received with deep emotion the welcome " intelligence of the fall of Sevastopol, with profound gratitude " to the Almighty who has vouchsafed this triumph to the " allied army.

" Her Majesty has commanded me to express to yourself, " and through you to her army, the pride with which she regards " this fresh instance of their heroism. The Queen congratulates " her troops on the triumphant issue of their protracted siege, " and thanks them for the cheerfulness and fortitude with " which they have encountered its toils, and the valour which " has led to its termination. The Queen deeply laments " that this success is not without its alloy in the heavy losses " which have been sustained, and whilst she rejoices in the " victory, Her Majesty deeply sympathises with the noble " sufferers in their country's cause.

" You will be pleased to congratulate General Pelissier, " in Her Majesty's name, on the brilliant success of the assault " in the Malakoff, which proves the irresistible force, as well as " indomitable courage, of our brave allies."

The loss of the regiment from the 16th December, 1854, to 8th September, 1855, was one officer (Captain J. L. Croker), one sergeant, and 34 rank and file killed; five officers (Lieutenants Boyd, Williams, Tompson, Swire, Parker), eight sergeants, and 120 rank and file wounded.

On the 4th October, the regiment, in brigade with the 20th, 21st, 57th, and 63rd regiments, formed a part of the expeditionary force sent against Kinburn, which however met with no opposition, the fortress capitulating after a heavy cannonading from the combined fleet. Brevet-Major Gordon was again mentioned in despatches on the termination of this successful expedition.

1856

On the 29th January, Major Gordon gave over command of the regiment to Lieut.-Colonel Cole, C.B., on the latter rejoining from sick leave.

On the 7th February, the regiment was inspected by Major-General R. Garrett, commanding 4th Division (851 rank and file on parade), and reported on "in excellent "order, as to interior economy and discipline," with about 320 men in it of five years' service and about 130 over 30 years of age.

From this time to the 28th February, the regiment was occasionally under heavy fire from the north shore fleets and batteries. Fort Nicholas was blown up on the 4th February, the docks having been destroyed on the 22nd December and 11th January, and the Russian barracks (commonly called the White Buildings) were blown up on the 28th February.

On the 29th February, firing was discontinued by general order on both sides, and, two days later, an armistice was concluded between the hostile forces.

On the 16th March, the whole of the British and French armies were reviewed by the Commander-in-Chief, on the hills near the monastery of St. George, but the day being bitterly cold, with a high wind, manœuvring beyond their line of march, to and from the camp, was given up.

On the 1st April, the new pattern double-breasted tunics were issued to the men.

The system of the supply of clothing by colonels of regiments (in accordance with the Clothing Regulations, issued in the time of Queen Anne, 14th January, 1707) had held good up to this date, and the colonels had already arranged for the annual supply for 1856–57. Consequently the clothing supply for 1857–58 was the first undertaken under the new system, which consisted of supply by contract. The supply of clothing for the foot guards, infantry, and cavalry had, up to this period, been a close monopoly in the hands of a few firms, and great opposition was made by these firms to the abolition of the monopoly. For the clothing required during the year 1857–58 the first contracts under the new system were made in 1856.

On the 17th April, the whole army was reviewed before the Russian general Ludyers, when some of the most distinguished of the Russian, French, and Sardinian generals were present. It was estimated that there were 30,000 British troops reviewed.

The following was the effective strength of the corps on its arrival in the Crimea, with the arrivals and casualties which took place during its service in that country :—

	Officers.	Men.		Officers.	Men.
Strength on arrival in the Crimea, 16th December, 1854	23	719	Killed, died and drowned	3	199
Joined from England, or Malta	52	589	Transferred	—	33
Transfers received ..	—	9	Sent home invalided ..	29	217
			Deserted	—	1
			Promoted to commissions	—	2
			Embarked for Quebec ..	29	865
			Sent home as supernumeraries	14	—
Total	75	1,317	Total	75	1,317

On the 23rd April, the regiment received orders to be held in readiness to embark for Canada, and on the 8th May marched to Balaklava.

Storming of Ghuznee, 1839 (The 17th planting their colours on the Citadel.)

(From an old print.)

THE COLOURS, 1848.

Plate 2.

PRIVATE SOLDIER.
1742.

Plate 3.

GRENADIER,
1751.

Plate 4.

GRENADIER.
1768.

Officer 1800.

Officer, 1814.

DRUM-MAJOR, 1830.

BANDSMAN (WITH SERPENT), 1830.

SERGEANT, GRENADIER COMPANY, 1830.

BUGLER, LIGHT COMPANY. (SUMMER UNDRESS), 1830.

Officer, Grenadier Company, 1831–45.	Officer, Battalion Company, 1830. (*Summer Dress.*)	Officer, Light Company, 1831–45.

OFFICER,
1834.

OFFICER AND PRIVATE,
1846.

Plate 10.

Pioneer, 1861—67. Private, 1861—70. Colour-Sergt, 1869. Private, Walking-out Dress, 1861—70. Bandsman, 1861—71. Officer (Levee Dress), 1871. Private (Marching Order), 1903.

THE COLOURS, 1910.

CHAPTER XI.

CANADA, HOME SERVICE, NOVA SCOTIA, JAMAICA, AND CANADA WEST.

1856–1867.

1856

Canada. 1865.

ON the same date the regiment embarked on board the steam transport " Robert Lowe " for Malta, and whilst in the Black Sea a fire broke out on board, which however was extinguished without any serious consequences. Arriving on the 17th May, the regiment was transhipped the same afternoon to H.M.S. " Vulcan," and having been completed to the full complement from the reserve companies at Malta, sailed for Gibraltar on the 18th, where they arrived on the 24th, and on the 30th May sailed for Quebec, but put back on the 1st June on account of an accident to the machinery.

On the 6th June Major Gordon was promoted to a brevet lieut.-colonelcy.

The regiment left Gibraltar on the 12th June, after the ship had put back a second time for repairs, and arrived at Quebec on the 20th July, having stopped at Halifax a week and at Pictou two days for further repairs. On landing, on the following day, the colours of the regiment were decorated with garlands of flowers by ladies of Quebec, and on the march to the citadel were carried by Lieutenants A. P. Traherne and A. H. Utterson, both of the light company.

On the landing-stage the regiment was received by the Mayor and Corporation, the whole of the garrison, and a large concourse of inhabitants, when the Mayor read the following address to Lieut.-Colonel Gordon, at the head of the regiment :—

" Allow me, on behalf of the Councillors and inhabitants " of the city of Quebec, to welcome you and the officers,

" non-commissioned officers, and men of the regiment under " your command on your arrival amongst us. The great " distances which you have traversed, the brevity and glorious " termination of the war in which you were engaged, and " your prosperous voyage, all combine to demonstrate that " bravery, genius, and devotion are the characteristic traits " of the great empire to which we belong, and of the age in " which we live.

" When we think of the successful siege of Sevastopol, " the greatest perhaps recorded in history—a siege which " collected together more than four hundred thousand men " from vast distances, the largest fleets which had yet covered " the seas; when we read the glorious history of the glorious " victories of Alma, Balaklava, Inkerman, Eupatoria, and, " finally, the successful storming of the great fortress of the " East, we feel ourselves carried to the celebrated historical " era of the West, stopping the invasions from the East in " the 15th century.

" It is needless to describe the feelings that arise at the " presence of the brave soldiers who took part in the late war, " and whose breasts bear the honourable testimonials of the " manner in which they discharged their duty on the field of " battle. We have already appreciated their heroic conduct, " and we shall ever be proud to have such noble models amongst " us to lead us in the road of honour.

" Permit me again, Sir, in the name of the inhabitants of " Quebec, to welcome you and the regiment under your com- " mand to this city, and believe me we shall at all times be " happy to do everything in our power to render agreeable " your sojourn amongst us."

To this Lieut.-Colonel Gordon made the following reply:—

" In the name of the officers, non-commissioned officers, " and men of the regiment under my command, I beg to " return you, the Councillors and inhabitants of the city of " Quebec, my most sincere acknowledgments for the address " you have done us the honour to present to us. I can assure " you it touches us deeply. The sentiments of kindness " and congratulation you have expressed towards us on our " arrival in your city are most gratifying, and fill us with " emotion.

"To have had the good fortune to form part of the "army which fought in the Crimea will ever be to us a source "of congratulation, which in after life we will look back to "with a feeling of pride, and which must always be increased "by the recollection of the brave allies who fought with us. "It is with the utmost satisfaction we find ourselves quartered "in your loyal city, and we anticipate passing a lengthened "and happy time amongst you.

"Again I beg, on behalf of the 17th Regiment, to return "you my most grateful thanks for the kind and cordial welcome "with which you have received us."

A procession was then formed, which accompanied the regiment through the town to the citadel. Numerous floral arches had been erected *en route,* and flowers were showered from the windows of houses as it marched to its quarters. This was followed, on the 30th July, by a dinner given by the citizens of Quebec to the non-commissioned officers and men, and at night they entertained the officers at a ball.

On the 20th August, the following were presented with the war medal, granted by the Emperor of the French, as a reward for their distinguished services during the siege of Sevastopol, viz.: Sergeants David Borrett, Charles Collins; Corporal Philip Smith, Privates John Davis, Richard Hogan, John Lawless, and Benjamin Vaughan.

At the annual inspection, at Quebec, by Major-General R. Garrett, K.H., on the 8th October (672 rank and file under arms), fault was found that there had been 50 courts-martial in the regiment in the month of September, chiefly for drunkenness.

On the 13th October, the establishment of the regiment was reduced to eight service and four depot companies, and on the 29th of the month a readjustment placed it at: 39 officers, 6 staff, 67 sergeants, 25 drummers, and 1,000 rank and file; of the latter, the service companies were to total 800 and the depot companies 200 rank and file.

In consequence of the reduction of four companies, Captains Boyd, Swire, Tompson, and Dyer were placed on half-pay on the 10th November, the supernumerary subalterns being retained on full pay.

1857

On the 1st August, at a garrison parade at Quebec, the Victoria Cross was presented to Lance-Sergeant Philip Smith, for distinguished bravery on the 18th June, 1855.

On the 2nd September, the regiment marched from the citadel to the Jesuit Barracks, and, on the 10th, Colour-Sergeant Henry Gibson was decorated with a war medal, given by the King of Sardinia, as a reward for his services in the Crimea, a similar decoration being awarded to Brevet-Major M'Kinstry, serving with the depot companies.

In the Inspection Report of the regiment, which took place on the 5th October this year, the officers' charge for messing is shown as 2s. daily except in summer, when the messing charge to each was 32s. a month.

Although the regiment was not in India during the mutiny, Captain Roger Swire, of the 17th (who had been severely wounded in the Crimea), took part in it. In General Windham's despatch to General Sir C. Campbell, Commander-in-Chief (dated Cawnpore, 13th November, 1857), the former, describing his actions of the 26th to 29th November against the mutineers of the Gwalior contingent, writes :—

"I have no staff of my own except Captain Roger Swire, "of the 17th Foot, my aide-de-camp, who has behaved with "his usual zeal and courage."*

On the 3rd November, three companies of the regiment, under command of Major Ruttledge, proceeded by rail to Montreal to be stationed.

1858

On the 7th May, the headquarters and seven companies under Lieut.-Colonel Cole, C.B., proceeded by river steamer to Montreal.

The establishment of the regiment had now been increased from eight to ten service companies.

In a Battalion Morning Order, dated Montreal, 15th June, Colonel Cole desired to place on the records of the regiment, the feelings of heartfelt sorrow with which he announced the death of Lieutenant W. H. J. Disbrowe, whilst employed as

* "History of the Indian Mutiny," G. W. Forrest, C.I.E. Vol. II.

aide-de-camp to Lieut.-General Sir W. Eyre, K.C.B., which occurred at Sorel on the 12th June. Colonel Cole bore testimony to Lieutenant Disbrowe's merits as an officer of rare abilities, who had served with conspicuous gallantry at Sevastopol whilst holding the more important and arduous duties of adjutant, and that he died beloved and universally lamented by the whole regiment, by none more so than his commanding officer, whose comrade he had been as well as intimate and valued friend (see Plate 9).

On the 28th June, at a garrison parade, the following officers received the insignia of the Order of the 5th Class of the Medjidie, granted by the Sultan of Turkey for their distinguished services in the Crimea, viz.: Colonel Cole, C.B., Lieut.-Colonel Gordon, Major Ruttledge, Major Brice, Captains Smyth and Dyer.

Formation of the Second Battalion.

The rebellion in India in the years 1857–58 (during which many European families were massacred), induced the Home Government to largely augment the army. Twenty-five new battalions being now added to the first 25 regiments of Infantry of the Line, the 17th "Leicestershire" Regiment again received a second battalion. Flank companies now became abolished, and the 2nd Battalion 17th was raised without them. The Adjutant-General's letter, dated 24th March, 1858, directed its formation by Lieut.-Colonel Hugh Crofton, who had been on half-pay from wounds received at Inkerman when serving with the 20th Regiment.

The nucleus of the battalion was formed at Limerick, where the 17th Depot Battalion (of which the 1st Battalion 17th formed part) was then stationed, and on the 23rd April embarked at Cork about 120 strong for Plymouth, being on arrival quartered at Maker Heights and Picklecomb Fort. It was here joined by Lieut.-Colonel Crofton, and such progress was made with its formation that, at the first half-yearly inspection in June following, the battalion mustered 20 officers and 369 rank and file, fully clothed and equipped, a transfer of 20 sergeants, 12 corporals, 5 drummers and 168 privates having been made from the 1st Battalion.

The Army List for that month shows the following officers posted to it, viz.: Lieut.-Colonel Crofton, Majors M'Kinstry and Rawstorne, Captains Boyd, Tompson, Fitzgerald, Davidson, Walton, McNair, Hunter, and Stuart, Lieutenants Hartwell, Webber, Colquhoun, Ensigns Wedderburn, Braddell, Ross (adjutant), Bros, Burnett, Caird, Weir, Deane, Elgin, Paymaster Smith, and Quartermaster Faulkner.

The enlistment at this time was for ten years—or twelve in case of war—the recruit received a bounty of £2, which was increased to £3 later on; the pay of the private soldier was 1s. per day, with an allowance of one penny per day termed "liquor money," out of which he paid 8d. per day messing, washing 1s. 3d. per month, sheet washing 2d., hair cutting 1d., and barrack damages an average of 4d. per month, the soldier getting the balance by daily payments, which usually came to about 4d. per day if no other necessaries were required.

On the 1st August, the 2nd Battalion moved to the citadel at Plymouth.

1859

On the 3rd February, the battalion received its colours from Lady Vivian on Mount Wise parade ground, and was then over 700 strong; average height, 5 feet to 5 feet 6½ inches. On this occasion the whole garrison paraded, when a hollow square was formed, the 2nd Battalion 17th being in line facing Government House. The right face of the square was composed of Royal Artillery and the 96th Regiment, and the left face, of the Royal Marines and 2nd Warwick Militia, the colours being received by Ensigns Bros and Deane. In the evening the officers gave a colour ball at St. George's Hall, Stonehouse.

On the 12th May, the battalion, under command of Colonel Crofton, was inspected at Devonport by H.R.H. the Duke of Cambridge, who complimented the commanding officer on its smart, clean appearance and general efficiency.

On the 1st July, the 2nd Battalion left Plymouth for Aldershot, travelling all night, and arrived at 6 a.m. next morning at the North Camp, where it remained a week under canvas, prior to moving to the east block of the Permanent Barracks, South Camp, when it was posted to the 2nd Brigade.

On the 5th October, the 1st Battalion proceeded by the Grand Trunk Railway from Montreal to Quebec.

On the 16th December, Colonel Crofton exchanged with Colonel Francklyn, and took over command of the Depot Battalion at Preston.

1860

On the 7th August, Major-General Sir Richard (afterwards Lord) Airey, K.C.B., was appointed Colonel of the regiment vice Lieut.-General T. J. Wemyss, C.B., deceased.

On the 29th, the 2nd Battalion left Aldershot for Shorncliffe.

On the 31st August, the establishment of the 1st Battalion was augmented by one staff-sergeant, styled "Sergeant Instructor of Musketry."

1861

2nd Battn. Ireland. 1861.

On the 22nd April, the 2nd Battalion proceeded to Ireland, and was quartered at the Richmond Barracks, Dublin, for six weeks, when it marched to the Curragh Camp.

On the 17th May, Major Alexander M'Kinstry succeeded to the command of it, on Colonel Francklyn's retirement from the service.

In December, this year, what was known as the "Trent affair" made a considerable commotion at home for a while, consequent on the representation of Messrs. Slidell and Mason, of the Southern Confederacy, who had been made prisoners from the Royal Mail Steamship "Trent" by Captain Wilkes of the United States warship "San Jacinto."

2nd Battn. Nova Scotia. Dec., 1861.

It having been deemed advisable to increase our troops in British North America, the 2nd Battalion 17th received orders to embark for Halifax, Nova Scotia, and the head-quarters and six companies embarked at Cork on the 27th December, in the steamship "Cleopatra," arriving after a good passage of eleven days.

1862

Four companies of the left wing (under Major Colthurst) embarked on the 1st January at Cork in the steamship "Mauritius," and, after an extremely rough passage of four weeks, arrived there on the 28th.

The above difficulty having been settled without hostilities, the battalion was enabled to settle down in its new quarters in the Wellington Barracks, the other regiment in the garrison being the 2nd Battalion 16th. It was this year authorised that companies were to be distinguished by a letter and to stand on parade according to the seniority of their captains.

1863

The garrison duties in Halifax, furnished by the two regiments, included the Main Guard and guards at the Main Dockyard, Ordnance, Magazine, General's House, and General Hospital, all of which came very heavy in the summer, when detachments had to be supplied to George's Island, York Redoubt and other places, besides Chobham Camp, where musketry was carried out, and in addition there were outposts (generally a sergeant and four privates) at Dartmouth, North West Arm, and Chezzetcook.

At this time the American Civil War was in progress, and tempting bounties being freely offered for trained men to join the United States army, the above posts were established in order to prevent desertions from the army and navy of the garrison.

1864

Throughout this year both battalions of the regiment remained at their respective stations.

1865

On the 1st April, the establishment of the 1st Battalion was reduced by fifty privates.

st attn. ngland. 865.

On the 20th May, the 1st Battalion embarked at Quebec for England, and arrived at Aldershot on the 5th June.

Prior to its departure from Canada, the following address was presented to it :—

" We, the undersigned magistrates for the city and district " of Quebec, have great pleasure, on the occasion of your " departure from amongst us, in conveying to the officers, non- " commissioned officers, and men of the regiment the thanks " of this community for your zealous support of the laws, "and the willingness with which you have, on all occasions, "afforded your aid when required, to assist the citizens or the "constituted authorities, during a residence of nearly nine "years among us. We have also great satisfaction in acknow- "ledging, that no regiment could have done more than yours "in cultivating and maintaining a good understanding between "yourselves and the citizens, by whom you will be agreeably "remembered. The duties performed among us, though less "brilliant than those of many a battle field, have nevertheless "been no less useful or conducive of good results.

"The social relations which existed between yourselves "and our citizens, which individual exertions have done so "much to render pleasant, will not be readily forgotten by us.

"Finally, we wish you a pleasant and prosperous voyage "across the Atlantic, and such success in other climes as strict "discipline and good conduct alone can ensure, and that you "may meet as many and sincere friends as you leave behind "you here."

(Signed by 38 Magistrates and Justices of the Peace.)

To this Colonel Gordon replied :—

"On the part of the officers, non-commissioned officers "and privates of the 17th Regiment, I thank you for your "address.

"To find that our endeavours to support the laws and "constituted authorities, during the very long time we have "been amongst you, have been appreciated is most gratifying "to us. In cultivating and maintaining a good under- "standing between the citizens of Quebec and ourselves, we "performed a most agreeable duty, and that our doing so "has been useful and conducive of good results is owing to the "kindness and other good qualities of the citizens.

"The social relations which have existed between them "and ourselves have been of a most happy and cordial nature. "They are impressed on our hearts, and will never be for- "gotten. We thank you for your good wishes for a prosperous "voyage to England and success in our future career, and we "assure you that no change of time or clime will efface from "our memory the many and sincere friends we leave in "Quebec."

nd
Battn.
Jamaica.
1865.

On the 28th October, the 2nd Battalion proceeded in H.M.S. "Duncan" (flagship of Admiral Sir James Hope, K.C.B.) and H.M.S. "Sphinx," to Jamaica, with a view to quelling a rebellion amongst the negroes in that island, but its services on arrival not being required, the battalion returned to Halifax in H.M.S. "Galatea" and H.M.S. "Aurora," arriving on the 30th December. Letters were received through Admiral Sir James Hope, from the captains of these ships, eulogising the exemplary conduct of the men, against whom there was not a single complaint. The letters testified to their cheerfulness, notwithstanding the unavoidable discomfort borne by so many more than the ship's company.

Previous to this, a strike had taken place on the island of Cape Breton, to which port a strong detachment of the regiment was sent under Major Heigham, and the colliers were soon induced to return to their work.

1866

After the Civil War in America, a number of discharged troops, composed of English, Irish, Scotch, Germans, French, Dutch, &c., from the service of the United States, formed themselves into a body for the invasion of Canada, and, calling themselves Fenians, assembled on the New Brunswick frontier, threatening to invade the British provinces.

On the 17th April, the 2nd Battalion again embarked at Halifax on H.M.S. "Duncan" for conveyance to St. Andrews, New Brunswick, accompanied by a battery R.A., detachment of Engineers, and two corvettes of the fleet. General Sir Hastings Doyle, with his staff, accompanied this expedition, and took command of the troops at St. Andrews, but the United States government having sent a force to co-operate with the British, the invaders dispersed and the field force re-embarked in H.M.'s troopship "Simoom" for Halifax.

News having reached Halifax later, that the Fenians had crossed the border at Fort Erie, and had fired on the volunteers, the battalion, on the 4th June, embarked with light baggage for that province, the headquarters and seven companies (for the third time) on H.M.S. "Duncan," and three companies on H.M.S. "Wolverine" for Quebec, the headquarters proceeding to Montreal.

After a month of uncertainty, the battalion proceeded by river steamboats to Toronto, Canada West, where it arrived on the 12th July, and encamped at the old fort until quarters were ready at the Parliament buildings.

2nd Battn. Canada West, 1866.

The battalion now furnished an officer's guard daily over the Fenian prisoners at the old gaol, Toronto, 120 of whom, from Buffalo, had been made prisoners in their attack on Fort Erie.

To those troops who took part in suppressing the Fenian raid, a medal with clasp was issued in 1899, but owing to the length of time that had elapsed (33 years) it had to be specially applied for by both officers and men who were entitled to it.

On the 30th August, the 1st Battalion proceeded from Aldershot to Devonport.

1867

In January, the Snider (converted Enfield) rifle was issued to the 2nd Battalion in Canada.

On the 17th April, the 1st Battalion embarked at Devonport for Ireland, arriving at Cork on the 20th, and on the 24th May the regiment proceeded to Kilkenny, furnishing detachments to several out-stations.

1st Battn. Ireland, 1867.

On the 24th July, the left wing of the 2nd Battalion proceeded by rail from Toronto to Brantford, C.W. (in relief of the 2nd Battalion 7th Fusiliers, ordered home), a wild part of the country within a few miles of an Indian settlement, and remained there until the 4th September, when it returned to Toronto, on relief by the 69th Regiment from England.

CHAPTER XII.

HOME SERVICE, EAST INDIES, AFGHANISTAN, AND EAST INDIES.

1868–1881.

1868

ON the 27th January, the 1st Battalion moved from Kilkenny to the Curragh Camp, and from the 1st April, the establishment was augmented by 20 privates.

From the 1st May, Major-General John Grattan, C.B., was appointed Colonel of the regiment, vice Lieut.-General Sir Richard Airey, K.C.B., appointed to the colonelcy of the 7th Fusiliers.

[2]nd [B]attn. [I]reland, [1]868.

On the 13th May, the 2nd Battalion embarked at Toronto, by wings, for Montreal and Quebec, and the headquarters and six companies transhipped later to the steamship " Moravian " *en route* for Dublin, where they arrived on the 31st May and were quartered at the Royal Barracks. Four companies had been left behind at Kingston, and returning in H.M.S. " Himalaya," rejoined in Dublin in July.

On the 9th October, the 1st Battalion moved from the Curragh to Newry, with detachments at Enniskillen and other out-stations.

On the 13th October, the 2nd Battalion moved to Mullingar with detachments at Boyle and Sligo, and later furnishing detachments to Dunmore and Esker. At one period during the Irish Elections this year, this battalion was split up into eleven different detachments, marching all over the country.

1869

From the 1st April, the establishment of the 1st Battalion was directed to be: three field officers, 12 captains, 14 lieu-

Colonel W. Croker, C.B.

From an Oil Painting.

tenants, 10 ensigns, 5 staff, 59 sergeants, 50 corporals, 25 drummers, 800 privates.

On the 10th April, the 2nd Battalion left Mullingar *en route* for the Channel Islands, via Dublin and Kingstown, the headquarters and five companies proceeding in H.M.S. "Orontes" to Jersey, two companies to Guernsey and three to Alderney, the last-named detachment being relieved every six months.

2nd Battn. Channel Islands, 1869.

On the 25th August, the 1st Battalion moved to the Richmond Barracks, Dublin, prior to embarking for India, and on the 17th December was sent to Londonderry in aid of the civil power, and returned to Dublin on the 20th.

1870

On the 12th January, the 1st Battalion, under command of Lieut.-Colonel A. H. Cobbe (who had exchanged from the 20th) embarked at Cork, for conveyance to Bombay (total strength, officers and men, 827), where it landed, on the 16th February, for the third time in the 19th century, and proceeded to Lucknow. Prior to embarkation the number of service companies was reduced from ten to eight, without altering the numerical strength except in officers, colour-sergeants, and drummers for the two reduced companies.

1st Battn. East Indies, 1870.

On the 26th July, the 2nd Battalion left the Channel Islands for Aldershot, via Portsmouth, and was encamped on Cove Common, then moving into huts at the North Camp, and later to the West Infantry Barracks, South Camp, joining the 2nd Brigade.

2nd Battn. England 1870.

1871

On the 30th April, Major-General W. R. Faber, C.B., was appointed Colonel of the regiment, vice Major-General J. Grattan, C.B., deceased.

By Royal Warrant, dated 30th October, purchase in the army and the ranks of Ensign and Cornet were abolished, that of Sub-Lieutenant being substituted for them.

On Colonel M'Kinstry's retirement from the service this year, Colonel G. T. Brice succeeded to the command of the 2nd Battalion.

1872

On the 1st April, the establishment of the 1st Battalion received an increase of a Pioneer-Sergeant to the service companies.

After the manœuvres on Salisbury Plain this year, the 2nd Battalion moved to Plymouth, and took up quarters at the Citadel and Mill Bay Barracks, and were shortly after transferred to the Raglan Barracks.

1873

On the 29th November, the 1st Battalion left Lucknow by route march for Peshawur; whilst quartered at Lucknow it had lost one officer and 94 men by death, and immediately prior to marching out the regiment had been the greater part of November in four cholera camps with the loss of ten men.

1874

On the 2nd March, the 1st Battalion arrived at Peshawur and occupied barracks in the Left Infantry lines, facing the Khyber Pass, having been marching for three months and four days and covered 896 miles.

Throughout the hot weather the headquarters and two companies of the battalion, also the weak and sickly men, occupied the sanatorium of Cherat (periodical exchanges of men being made in July and September); and during the last four months of this year the battalion suffered much from the deadly fever and ague peculiar to the Peshawur Valley at that period, which was in a great measure subdued by being sent out to camp and by marching about the district.

On the 3rd November, the battalion re-occupied quarters at Peshawur.

nd attn. reland, 874.

On the 25th June, the 2nd Battalion embarked at Plymouth for Ireland and proceeded to the Curragh. Since 1872 the depots of the 1–17th and 45th Regiments had been attached to the 2nd Battalion, two companies of which, with these depots, now proceeded to Athlone, where the headquarters joined them in September, furnishing detachments to Castlebar, Ballina, Tuam, Ballaghaderin, Ballinrobe, and Newport.

1875

On the 11th January, the 1st Battalion was reviewed with the other troops in garrison at Peshawur by Lieut.-General Sir C. Staveley, Commander-in-Chief of the Bombay Army.

After the parade Brigadier-General Wilson, addressing the troops, said that His Excellency had desired him to say that he " was extremely pleased with the appearance and " marching of the troops, especially that of the 1st Battalion " 17th, which regiment he was glad to see still maintained " its renown for marching past."

On the 28th January, the battalion was inspected by Lord Napier of Magdala, Commander-in-Chief in India, who at the conclusion of the inspection, expressed himself to Colonel Cobbe, as stated in the following Regimental Order dated 30th January, 1875 :—

" The Commanding Officer experiences much pleasure " in announcing to the regiment, that the Right Honourable " the Commander-in-Chief was pleased to express himself " highly satisfied with the state of the regiment at his recent " visit to Peshawur. His Excellency was pleased to remark " most favourably on their soldier-like appearance, dress, " good conduct, and movements on parade, and to add that " the favourable opinion he had formed of the regiment at his " former inspection, at Lucknow in 1870, was quite un- " changed."

In May, the 2nd Battalion returned to the Curragh from Athlone and out-stations.

1876

2nd Battn. East Indies, 1876.

On the 25th March, the 1st Battalion proceeded by route march to the Murree Hills, and, prior to its leaving Peshawur for Rawal Pindi, the following District Order was published :—

" The Brigadier-General cannot permit the 1st Battalion " 17th Regiment to leave this command, without placing on " record the high opinion he has formed of the regiment. " Brigadier-General Ross much regrets the departure from the " district of such a fine and soldier-like regiment."

On arrival at Rawal Pindi, the battalion proceeded to four of the principal gullies in the Murree Hills, viz., Kuldanna, Changla Gali, Kalabagh and Bara Gali, which it occupied in four double-company detachments throughout the summer months, returning in the autumn to a standing camp at Rawal Pindi until the following spring.

In June, the 2nd Battalion left the Curragh for Templemore.

On the 5th October, the battalion embarked at Queenstown, Ireland, for conveyance to Bombay, where it arrived on the 7th November, under command of Colonel G. T. Brice, and proceeded to Mhow with a detachment at Indore. The depot was left at Templemore and later proceeded to Newport, Monmouth.

The rank of 2nd Lieutenant was substituted for that of Sub-Lieutenant in the Royal Warrant of the 30th October, 1876.

1877

On the 1st January, Her Majesty Queen Victoria was proclaimed "Empress of India." The 1st Battalion at Rawal Pindi paraded with all the troops in garrison to hear the Proclamation read, and to celebrate the occasion. A Royal Salute of 101 guns and a *feu-de-joie* were fired, and the parade terminated with three cheers for Her Majesty the Queen and Empress of India. To mark the event the Government presented a day's pay to every soldier, British and native, in India. A silver commemorative medal was also given to every regiment, to be presented to a selected soldier. The medal given to the battalion was presented to Sergeant-Major George Williamson.

In the spring the battalion proceeded to the Murree Hills as in the previous year, with a change of companies, returning to a standing camp at Rawal Pindi for the winter months.

The Martini Henry rifle, having been adopted as the weapon for the infantry of the army, was issued to the 1st Battalion on the 5th June.

1878

The 1st Battalion marched to the Murree Hills in the spring of this year, as in the two previous years, with headquarters at Kuldanna.

In October, the 2nd Battalion proceeded from Mhow to Nusseerabad and Neemuch, the headquarters arriving at Nusseerabad on the 1st November.

THE SECOND AFGHAN WAR.

When, on the 21st September, the advance of a mission (under Sir Neville Chamberlain) from Lord Lytton, Viceroy of India, to Shere Ali, Ameer of Afghanistan, was refused permission to proceed beyond Ali Masjid, an ultimatum was sent to the Amir, embodying an explicit declaration that, unless a satisfactory reply was received by the 20th November, the English troops would cross the frontier. When the specified date arrived, the Amir having failed to reply, the order was given for the British troops to advance into Afghanistan.

The number of troops to operate, advancing in three columns, was estimated as follows, with names of commanders :—

Peshawur Valley Army, 10,000 men, with 30 guns, Lieut.-General Sir S. Browne, V.C., &c. ; Kurram Column, 5,550 men, with 24 guns, Major-General Roberts, V.C., &c. ; Quetta Army, 6,250 men, with 18 guns, Major-General Biddulph, in command, until the 15th December, when General Donald Stewart re-inforced with the Mooltan Division and assumed command of the Quetta Force.

The 1st Battalion 17th was one of the first corps ordered to prepare for service. Preparations were at once commenced. " Khaki " (dust colour) clothing was served out, the pipe-clay washed out of the belts and the further use of pipe-clay and blacking discontinued, the boots being greased and pouches and scabbards covered with khaki. The clothing was made loose to permit its being worn over serge or cloth uniform if necessary.

Early in October, the battalion proceeded from the Murree Hills to Rawal Pindi, and, on the 14th, marched *en route* for Peshawur, having exchanged European privates' tents for sepoys' " pals."

On the same date, Colonel Cobbe was appointed Brigadier to a brigade with the Kurram Column, with Lieutenant Reader as his orderly officer.

Major (Brevet Lieut.-Colonel) Tompson accordingly assumed command of the battalion, which was posted to the 2nd Brigade, 1st Division, the remaining troops of the brigade being the Guides Infantry, 1st Sikhs, and a mountain battery

R.A., the brigade being under the command of Brigadier-General Tytler, V.C., C.B.

t
ttn.
ghanis-
n,
78.

On the 16th November, the Colours of the battalion were stored in the Peshawur arsenal, in accordance with instructions from the Lieut.-General commanding, as is usual with British regiments on field service in the hills, and by the 19th November, the battalion, with brigade headquarters, had moved to Jumrood on the frontier, and was joined by the other regiments of the brigade.

With the object in view of intercepting the retreat of the garrison of Ali Masjid, the 2nd Brigade, in conjunction with the 1st (both taking different circuitous routes), was directed to make a flank march in the mountains on the fortress of Ali Masjid, and at 6 p.m. on the 20th November, the 1st Battalion 17th (strength, 634 of all ranks), marching from Jumrood, was the first European regiment to cross the frontier, Sir S. Browne having determined to make a frontal attack with his 3rd and 4th Brigades.

Surgeon-General Evatt, C.B., writes of the regiment :—

" If it be not invidious, the palm for physical fitness and " complete efficiency on the old long service lines might be " given to Colonel Tompson's battalion of the 17th Foot. " These physically fine men, this regiment of grenadiers, had " just come down from the Murree Hills. No regiment ever " entered the field in finer condition, and I can never forget " the sight, as this battalion, in the pink of physical efficiency, " marched out of Jumrood Camp, some eight miles from the " mouth of the Khyber Pass on its perilous journey. It was " about the last of the long-service battalions of that army " which was just then disappearing before the short-service " system, and better specimens of that old *régime* could not " be seen than the men of the 17th, who, for weight and space " occupied per man, were probably 30 per cent. heavier and " broader than the young soldiers of to-day."

The following officers of the 1st Battalion crossed the frontier on the evening of 20th November :—

Lieut.-Colonel W. D. Tompson (commanding),
,, A. H. Utterson,
Brevet-Major C. M'Pherson,
,, J. O. Travers,
Captain and Local Major G. H. Turner, (Officiating Paymaster),

Captain W. Lonsdale,
,, J. H. Gamble,
,, C. F. W. Moir,
Lieutenant and Local Captain C. W. Vulliamy,
Lieutenant E. A. H. Webb,
,, G. J. Buller,
,, N. C. Wiseman,
,, C. W. Boddam,
,, R. J. G. Creed,
,, M. R. Hyslop,
2nd Lieutenant W. S. Stewart-Savile,
,, G. J. Younghusband,
Sub-Lieutenant W. Cook (89th Foot),
Lieutenant, Local Captain and Adjutant J. G. Anderson,
Quartermaster J. Fallon,
Surgeon-Major W. B. Ramsbotham.

The 2nd Brigade did not meet with any opposition in the course of its arduous march to the Khyber Pass, the march itself being principally celebrated for the difficulties encountered in the nature of the ground, which at times necessitated the whole column of about 1,800 men marching in single file, over slab rocks and steep rugged stones, through narrow gorges with frowning precipices (where the sky had to be looked for in the narrow break of the overhanging rocks above)—places, in fact, where only man or monkey could have ventured at all—whilst, to crown all, the day's cooked rations, which were being carried for the regiment by the commissariat, had (with a want of prevision which cannot be too severely blamed) been placed on bullocks instead of mules. It is supposed that they never reached the heights at all, but that, dismayed by the difficulties of the way, they simply returned in peace to Jumrood. It was principally on this account that the Guides and 1st Sikhs were sent on ahead, so that when the 17th arrived on the heights of the Khyber Pass (on the morning of the 22nd November) all was over. Ali Masjid had fallen, and the Guides and 1st Sikhs had captured over 200 prisoners retreating from the fortress, and had also made short work of some of the Afghan cavalry retreating up the Pass. On the same evening, rations, in the form of seventeen sheep, which had been captured in the Pass by the picquet of the Guides, were handed over, by the General's order, to Colonel Tompson for distribution to the companies of the 17th, and the sheep were at once killed, cooked and eaten.

The measured distance marched by the 2nd Brigade on this occasion was 25 miles, in the course of which they ascended to a height of nearly 5,000 feet by rough paths and over rocks and boulders, more adapted to goats than soldiers.

The works at Ali Masjid were found by the 3rd and 4th Brigades to be of great extent, and to have been most skilfully and formidably designed. Had they been manned by a sufficient force, and precautions taken against flanking movements, the position would have been almost impregnable, and a frontal attack alone could only have been successful at the cost of an immense loss of life.

On the 23rd, the main body of the Peshawur Valley Army marched 13 miles up the Khyber to Lundi Khana, and on the 24th advanced to Dakka, a large cantonment of the Afghan army situated in a circular basin, enclosed by hills and traversed by the Cabul River.

At Dakka, Sir S. Browne received the congratulations of Her Majesty the Queen, H.R.H. the Prince of Wales, and the Viceroy of India, on the complete success of his brilliant operations in the Khyber; in communicating this to the troops, he expressed his appreciation of the cheerfulness and soldier-like spirit with which they had worked.

The troops were now employed in reconnoitring, escorting convoys, and putting the fort into a serviceable condition.

News was here received that Colonel Cobbe of the 1st Battalion (Brigadier-General with the Kurram Valley Field Force) had been wounded on the 2nd December at the attack on the Peiwar Kotal.

Many cases of plundering now occurred along the Pass, in the rear, and attacks were made by the hill tribes upon outlying sections of the force left at Ali Masjid. A section of the Shinwarri tribe, called the Mirzan Khel, now gave some trouble, and cut off and killed four grass-cutters of the British force, so, at midnight, on the 9th December, an expedition was sent (the battalion furnishing 13 officers and 300 rank and file, under Lieut.-Colonel Tompson), to surprise their village, (known as Chenar), which was partially destroyed, without meeting with any resistance. The prompt punishment of this tribe produced a salutary effect.

The *Times* Correspondent, writing from Ali Khel on December 11th, says that "Brigadier-General Cobbe is "mending fast." The same correspondent says:—"The health "of the troops at Dakka is fairly good. The 1st Battalion "17th Regiment remains pre-eminent for good health and "fitness for service. They have only some eight men in "hospital, and the hard work of the mountains has harmed "them but little." Further on, when speaking of the night attack of December 9th, he writes:—"And here I must "again record my meed of praise, re-echoed on every side, as "to the splendid endurance of that gallant corps the 1st "Battalion 17th. Three hundred men of the regiment took "part in this operation, and were called on to perform in "eighteen hours a thirty-one mile march over the roughest of "ground. Yet they marched in at night apparently fresh "and smart as if returning from an hour's drill; and, on "enquiry, it appeared that only three men of that number "had availed themselves during the entire march of the "dandies which followed the column. Two of these had "fallen over rocks and received injuries necessitating their "being carried. Only one man had to be carried from "exhaustion. As yet this corps has only shown its powers "of endurance as a splendid marching regiment, but it will "be an evil day for the opposing Afghans when it has an "opportunity of displaying its fighting qualities."

On the 13th December, three companies of the regiment advanced to Basawul and, on the 16th, the 2nd Brigade was reconstructed as follows:

1st Battalion 17th Regiment 27th Punjab N.I. 45th Rattray's Sikhs	Under command of Brigadier-General Tytler, V.C., C.B.

Sir S. Browne, having halted some weeks at Dakka, marched for Jellalabad on the 17th December, which was entered by our troops without opposition on the 20th, and the inhabitants showed a friendly disposition.

General Tytler was left in command at Dakka. It was ascertained about this time that Shere Ali had left Cabul on the 13th December, and had fled into Turkestan, leaving the government in the hands of his son, Yakoob Khan.

Looting was still being carried on in the Khyber Pass, the chief offenders being the Zakka Khels, a section of the Afridi tribe, who had given us great trouble in the Pass, making constant raids on our convoys, cutting off and murdering stragglers, and cutting and carrying off telegraph wire. Their offences culminated on the evening of the 18th December, in a bold attempt to capture the camels of the 2nd Gurkhas whilst grazing near Ali Masjid.

On the 19th December, the two columns from Ali Masjid and Dakka (the whole under command of Lieut.-General Maude, V.C.) co-operated in an attack against this turbulent tribe, the Dakka column, commanded by Brigadier-General Tytler, consisting of 300 men 17th Foot, under Colonel Tompson; 250 of the 27th Punjab N.I.; 100 of the 45th Sikhs, and some Sappers and Miners.

This column left Dakka at mid-day on the 19th, in the lightest marching order, carrying two days' rations, and the men's great coats and one blanket per man on mules. At 4.30 p.m. it halted in a stony plain covered with high stunted bushes and long grass, and throwing out pickets, bivouacked for the night. Early next morning (20th) the column reached the village of Chenar (which had been destroyed by our troops on the 10th), and here the head men came out to meet General Tytler and accompanied the column as guides. On arrival at Sitsobi, the head men also came to make submission to the General.

Sitsobi was found to consist of five considerable villages, fortified with mud walls, flanked by towers, and situated on the lower slopes of the surrounding hills. At 12.30 p.m. the march was continued to the Sitsobi Pass, the ascent of which was about 1,200 feet and very laborious, and from its summit a fine view of the Bazar Valley was obtained. The column then made a very steep descent towards the chief village of Bazar, its situation being indicated by a large round tower, which the troops reached at about 4 p.m. The inhabitants having received warning of the attack, had deserted it, leaving behind, in their haste, their cattle and stores of provisions and forage. The troops bivouacked here on the night of the 20th, and opened communication with Brigadier-General Doran, in command of the Ali Masjid column, which also had been doing good work. At 8 a.m. on the 21st December, General Tytler rode into General Maude's camp to confer as to the measures to be taken for the punishment of the Zakka Khels.

He was directed to destroy, on his return to Dakka, the villages in the vicinity of his bivouac, and then march to Nikai and destroy it, Nikai being a village reported to be the home of some of the most desperate of the clan. The Ali Masjid column, receiving instructions to destroy the villages bordering on the Bara River, on its return to Ali Masjid, General Doran accordingly blew up and burned Wallai and some other towers and villages, reaching Ali Masjid before nightfall, after meeting with some slight opposition. General Tytler, after destroying the villages and towers of Bazar, marched to Nikai, which, like the other villages, was found to be deserted, and was quickly reduced to ashes.

The guides, having now mentioned a pass called the Tabbai Kotal as being a more direct road to Dakka, General Tytler resolved to return by it, every precaution being taken to guard against surprise, the advanced guard and flanking parties being strengthened, the column well closed up, and two companies furnishing a rear guard. No opposition was experienced, and as evening fell a copious spring of water was reached, while several grassy flats afforded good ground for a bivouac. Outlying picquets were posted for the night, which passed without any attack by the hillmen in the neighbourhood, but, as the rear guard was marching into camp a private of the 17th Foot was shot through the thigh. Early next morning the march towards the defile leading to Dakka was resumed, every precaution being taken, as before, by General Tytler, to ensure the safety of the column. At about 8.30 a.m. the main body moved off, each corps having its baggage with it. The usual tactics of the hillmen, in harassing a retiring force, now became evident, as the column had scarcely started before the enemy opened a brisk fire on the troops as they moved along the steep zigzag path leading to the Kotal (pass over a ridge). The troops were, however, so well covered by the dense foliage of the trees that there were no casualties on the march to the pass. The enemy's fire across the valley was replied to by the flanking fire of the 45th Sikhs at long range. On the summit of the pass being reached, a spirited attack was made by the half company of Sappers and Miners, supported by a detachment of the 27th N.I., on a party of the enemy discovered on a hill commanding the road. The latter were quickly driven from this position, and a Sapper was here shot through the arm. The main body then crossed the summit of the pass and commenced the descent, under a dropping fire; General Tytler, with his staff, remaining on the top of the pass until the last man of the column had reached it. It was about 11 a.m. when the companies of the 27th and 45th N.I. withdrew from their positions commanding the road, and followed the main body down the pass. The rear guard, consisting of two companies of the 17th Foot, and one of the 27th N.I., was directed to hold the summit and check the enemy until this movement was completed, and then continue the march. Hardly had the rear guard commenced its descent when the enemy, with increasing boldness, followed for some distance, keeping up a rapid but ill-directed fire, from which there were no casualties. In the meantime the main body, whilst passing through a very difficult gorge, was attacked by parties posted in narrow ravines, a man of the 17th Foot being wounded, and the man of the same regiment, who had been wounded on the previous day, was again shot, and died a few hours afterwards. Half a company of the 17th, on being despatched up a hill, quickly dislodged the enemy and silenced his fire. A party of the 45th Sikhs then manned the heights at this point, to ensure the safe passage of the rear guard, after which the march was unmolested, and on a muster being taken at the first halt, the casualties were found to be one man killed and seven wounded, and no baggage, animal, or any article of property was missing. At 2 p.m. the march to Dakka was resumed, the road being long, and in places very rough, so that it was after 9 p.m. when the advance guard arrived, but before midnight every man and animal was safe back in camp. The distance traversed on this day by the troops was estimated at 22 miles, exclusive of the detours of flanking parties and skirmishers on the hills.

On the following day General Tytler issued an order in the following terms:—

"The men throughout displayed gallantry, endurance, "and coolness, which elicited my warmest admiration. There "was no hurry; mules that had thrown their loads were "quietly re-loaded under fire; the small number of rounds

"expended (1,029) alone proves the entire absence of hurried
"firing, and the endurance of officers and men is sufficiently
"evident, when it is considered that the force was marching
"and fighting continuously, without food, from 8 in the morning
"till 11 at night, through a totally unknown and very difficult
"country."

Although the hoped-for surprise of the Zakka Khels had not been effected, it was believed that the expedition would have valuable results, in shattering their prestige as to the inaccessibility of their region, and in showing them that their mountain fastnesses could be reached, and the most difficult passes could be surmounted by our troops, so that until peace was made with us the hill tribes would never be able to consider themselves safe.

On the 6th and 18th December respectively, 2nd Lieut. Griffith and Brevet-Major Brind joined for duty.

On the 24th the battalion advanced to Basawul, with Brigade headquarters.

1879

A camp follower of General Tytler's force having been killed at Basawul, and the village, about eleven miles off, to which the murderer belonged, having been ascertained, a mixed party of infantry, comprising 100 men of the 17th and a few of the Guides Cavalry, under command of Lieut.-Colonel Utterson, was sent out at 1 a.m. on the 23rd January, accompanied by a Political Officer, and, surrounding the village early in the morning, the force captured 73 prisoners and about 400 head of sheep and cattle.

The attitude of the Afridis continuing very unsatisfactory, it was determined to send, towards the end of January, 1879, a considerable force into the Bazar Valley for the three-fold purpose of: (1) Punishing the inhabitants of certain villages in which both thieves and stolen property were hidden; (2) To convince them that we could, at any time, enter their country and exact reparation for offences; (3) To obtain an accurate survey of the district.

General Maude took command of the whole force, consisting of three columns, operating from Jumrood, Ali Masjid, and Basawul, all similarly apportioned in numbers. The

Basawul column, Brigadier-General Tytler commanding, consisted of Guides Cavalry, 32 men, 2 mountain guns R.A., the 1–17th Foot, 27th N.I., and 45th Sikhs, each 300 men; the 4th Battalion Rifle Brigade, and 4th Gurkhas, each 200 men, and Sappers and Miners, 42 men. The troops were supplied with provisions for several days, and were fully equipped for field service. The columns marched independently on the 24th and 25th January. On the 25th, several towers were destroyed, and, on the 26th, detachments marched on Boorg, the two towers of which were destroyed, the troops then proceeding to Choora. There was some firing on the rear guard near Aocha Tanga, and two men of the 2nd Gurkhas were wounded. During the nights of the 26th and 27th January, the camp was repeatedly fired into, and one man of the 25th Foot was killed, and two wounded. On the 27th, communication was opened by General Maude with Brigadier-General Tytler, at the Sitsobi Pass, and, on the same day, detachments of the 25th Foot, and 24th N.I. (of the Jumrood column) scoured the hills, and, in conjunction with the 13th Bengal Lancers, killed seven or eight of the enemy. The hills near Aocha Tanga were also scoured by detachments of the 51st Foot and 6th N.I. (of the Ali Masjid column). The Basawul column took part in the total destruction of the Chenar and Sitsobi villages, and in General Maude's reconnaissance in force (28th January) of the Bokhar Pass, leading to the Zakka Khel villages in Bara. The enemy resisted strenuously. The pass was found quite practicable for troops, and the retirement was effected steadily. Our loss, during the day, was one man, 4th Gurkhas, killed; Lieutenant Holmes, 45th Sikhs, a sergeant R.A., and one man of the 4th Gurkhas wounded. The loss of the enemy could not be ascertained, but was probably considerable.

The above action, in front of the Bokhar Pass, would have been followed by a further advance into the Bara Valley, had not General Maude, on 1st February, received a message from Sir S. Browne, stating that the Mohmands, and Bajawaris, incited probably by Yakoob Khan, were preparing to attack Dakka and Jellalabad, and requesting that General Tytler's column might be sent back at once. This request was urgent, and though, without this column, General Maude's force might still have been able to force its way to the Bara Valley, it would probably have been insufficient to meet the powerful resistance which certainly awaited it there, not only by the Afridis, but by other tribes, who, not having yet quarrelled with us, were known to be assembled in the Bara passes. Accordingly, the Political Officer was directed to hasten, as much as possible, an understanding with the Bara Zakka Khels, and, on the following day, he reported himself as satisfied with the terms the tribes had agreed to. At the same time, he considered a further occupation of the Bazar Valley as no longer advisable, on the ground of its probably leading to further acts of hostility. The return of the entire force was accordingly ordered by General Maude, the respective columns reaching their destinations on the 4th February. The result of this expedition was considered satisfactory, with a trifling loss on our side of 5 killed and 8 wounded, and the Zakka Khels had subsequently fulfilled their agreement, and shown a disposition to remain friendly, at any rate, for a time.

In the meantime the threatened hostilities by the Mohmands and Bajawaris, which had been anticipated by Sir S. Browne, took the form, on the 6th, of an attack by about 12,000 of them on the villages of a friendly chief in the Kama district.

On the 7th February, 300 rank and file of the battalion, with officers, marched from Basawul to co-operate with a column moving from Jellalabad, to attack and disperse them, but no enemy having been discovered, General Tytler returned to Basawul, and by the 11th all excitement in this district had passed away. On this occasion Lieutenant Creed rescued a soldier from drowning who was being carried down a rapid in the Cabul River, for which he (Lieutenant Creed) was afterwards awarded the bronze medal of the Royal Humane Society.

On the 9th February, Captain Lonsdale's company advanced to Barikab to garrison that post, and on the 19th, Lieutenant E. Webb, with 100 rank and file and 200 of the 45th Sikhs (under Lieutenant Barclay) marched to Girdi Khus to assist the Sappers and Miners in road making. During a period of inactivity for a while, the headquarters of the battalion were employed in escorting treasure and in road making. In the latter occupation, great praise was at all times bestowed on the regiment by the superintending engineer officers for its excellent work, notably on the companies under Captain Lonsdale and Lieutenant Webb, which (between the 20th February and 26th March) completed a new road by the Cabul River from Basawul to Ali Boghan. This road, 22 miles long, had been commenced early in February by two companies of Sappers, and the excellent work, on completion, resulted in an alternative road from Jellalabad to Dakka, leading past Chardeh, Lachipoora and Girdi Khus.

On the 18th March, 200 rank and file of the regiment, under Lieut.-Colonel Tompson, with a mixed force, formed an expedition under General Tytler, to obtain satisfaction for the attack on a survey party by Shinwarris, near Maidanak, on the Pesh Bolak Road. The usual penalties were exacted: three towers at Girdi and four others were blown up without any opposition from the Shinwarris, who agreed to pay the fines inflicted. The object of the expedition was thus satisfactorily obtained without bloodshed.

The 2nd Brigade marched on the morning of the 24th March to Barikab, *en route* to Jellalabad. Before advancing, however, Brigadier-General Tytler had determined to visit and punish certain villages south of Pesh Bolak, the inhabitants of which had fired on a foraging party under escort of the 27th P.N.I.

Accordingly at 1 a.m. the brigade vacated Basawul, and a mixed force, under command of the Brigadier-General (comprising 250 of the battalion under Lieut.-Colonel Utterson), moved off. With the cavalry the Brigadier pushed on quickly, and reaching two villages at dawn, about four miles from Pesh Bolak, he at once attacked, dismounting a portion of his Lancers as skirmishers. In half an hour the infantry and the guns came up, under Lieut.-Colonel Utterson, and advanced against the villages, whilst the cavalry, threading their way across a nullah, charged right through some of the enemy, who were collected on the plain beyond. The latter stood firm for a short time, but the lances were too much for them, and they fled to the hills, leaving about 60 killed and wounded. Meanwhile the infantry, with the guns, had carried the village of Deh Surak, in which they halted for breakfast while the Sappers mined several of the towers. At ten o'clock,

General Tytler began to withdraw, having blown up six or seven towers, and was followed as far as Pesh Bolak by large masses of Afridis, who kept up a continual fire and at times even demonstrated to charge. They were kept at a distance, however, by a free fire of shell and musketry until, after about three miles, the country becoming more open, the retirement was covered by the cavalry in alternate squadrons.

The affair was well conducted, our force having two killed and four wounded; the enemy, who were estimated at over 2,000, acknowledging to a loss of 180 killed.

Early on the 27th March, the battalion, including the detachments which had been recalled, marched from Barikab to Jellalabad with the 2nd Brigade.

Jellalabad was found to be a miserable town in a rapid state of decay, and the mud wall which surrounds it was much split in every direction. Its four gates also were hardly worthy of their name. The town was in the form of a parallelogram, with two main streets crossing in the centre, the bazaar being in the one which ran from east to west, and roofed by planks stretched across from shop to shop, and plastered on top with mud, the houses being wretchedly constructed of mud and wood, the streets ill-paved and rough, and the lanes from the bazaar narrow and dirty.

On the 29th, the brigade was inspected by the Lieut.-General commanding the division, who expressed himself highly pleased with the appearance of the men.

On the 1st April, the left half battalion, under Brevet-Major C. M'Pherson, started for Futtehabad with a force under command of Brigadier-General C. Gough, V.C., commanding the cavalry brigade, to reconnoitre the country in advance towards Gundamuk. Later on, the pickets having reported masses of men assembling close to the camp, General Gough at once went out against them (leaving 300 men to protect the camp), and found some 2,000 Khugianis holding a strong position about two miles off. Having signalled this to Jellalabad, the remainder of the 2nd Brigade (including the right half and headquarters of the battalion), with two guns, I/C. R.H.A., moved off on the night of the 2nd to support

General Gough. On arrival, however, at about 2 a.m. on the 3rd, this force found that the battle had been fought and the enemy routed. The cavalry of the Guides had suffered severely, with 32 casualties, which included their commander. The 10th Hussars had also suffered, and there were six casualties in the battalion, including Lieutenant Wiseman, who was killed whilst gallantly leading a charge. Lieutenant Wiseman behaved with great gallantry, personally capturing a standard in a hand-to-hand combat.

After thoroughly reconnoitring the country around and towards Gundamuk, the battalion, together with the cavalry brigade, advanced from Futtehabad on the 13th April, and was concentrated at Safed Sung (Gundamuk) on the following day, together with the headquarters and most of the remaining troops of the 1st Division. No further resistance was met with, while many of the Khugiani chiefs came in and tendered their submission.

Pending the result of the negotiations which were now being carried on between the Government of India and Yakoob Khan (who had been proclaimed Amir of Cabul on the flight and subsequently reported death of his father, the Amir Shere Ali), every preparation was made for a rapid advance on Cabul on a reduced scale of baggage.

On the 28th April, a detachment of the regiment, under Lieut.-Colonel Utterson, marched from Safed Sung to reinforce the covering party, about four miles on the road in advance. This party was withdrawn two days after. Whilst in advance, the detachment furnished a party which buried the bleached bones of the last remnant of the 44th Regiment, who were slaughtered whilst making their last stand during the disastrous retreat from Cabul in 1842. These remains were found on a red hill about six miles from Safed Sung, which was pointed out by the country people as being the spot where the last stand was made.

On the 8th May, Yakoob Khan, the Amir, with his Commander-in-Chief and a small infantry escort arrived at Safed Sung, and was received with Royal honours, the British force lining both sides of the road for four miles out of camp, and a guard of honour of one captain, two subalterns and 100 rank and file of the battalion being drawn up at his tent.

On the 26th May, a treaty between the Government of India and Yakoob Khan, Amir of Afghanistan, styled the "Treaty of Gundamuk," was made and signed.

On the 31st May, the troops paraded in review order for inspection by His Highness the Amir.

Orders having been now received for the immediate evacuation of Afghan soil by the British force, the 2nd Brigade, 1st Division, was detailed to form the rear guard, and to march from Safed Sung on the 7th June, and the battalion was, on arrival, to halt at Lundi Kotal (the new frontier as fixed by the treaty) and to form part of the Khyber brigade.

This was a very trying march, the heat being intense and cholera rife all along the line. The whole battalion was daily on baggage guard, as it formed the rear guard to the whole retiring force, and arrived at Lundi Kotal on the 14th June. The death roll was also considerable from fever, heat apoplexy and sunstroke. Some corps suffered very much (the 10th Hussars alone burying at Dakka 19 men in one grave), and the battalion lost five from cholera and one from remittent fever. The sick especially suffered greatly from the heat, and the immense amount of dust; metalled roads there were none, and the water supply was scanty beyond conception.

The following is a list of officers who joined the battalion for duty this year, with dates of field service, the 16th August being that on which the regiment arrived at Peshawur on its return to India :—

Captain Michel, 25th January to 16th August, 1879.

2nd Lieutenant Watson, 25th January to 16th June (died 16th June).

Lieutenant Allfrey, 1st February to 13th May (died 13th May).

Lieutenant Clarke, 26th March to 5th August.

Surgeon-Major McWatters, 20th April to 16th August.

Lieutenant (Instructor of Musketry) T. de Burgh, 25th April to 16th August.

Captain M. W. Brock, 20th July to 16th August.

2nd Lieutenant St. Quintin, 29th July to 16th August.

2nd Lieutenant G. H. P. Burne, 4th August to 16th August.

The casualties in the battalion from the 20th November, 1878, to the 16th August, 1879, included: one officer killed, ten rank and file killed and wounded, and three officers and 40 rank and file died from disease.

The following officers were mentioned in despatches on the termination of hostilities, viz.: Lieut.-Colonels W. D. Tompson and A. H. Utterson, Major F. S. S. Brind, and Captain and Adjutant J. G. Anderson.

The following Brigade Order was issued by Brigadier-General J. A. Tytler, C.B., V.C., under date, Lundi Kotal, Afghanistan, 20th June, 1879:—

"On the breaking up of the 2nd Brigade, 1st Division, "Peshawur Valley Field Force, Brigadier-General Tytler, "C.B., V.C., desires to offer his warmest thanks to Commanding "Officers, Officers, N.C.O.'s, and men of the brigade, for the "admirable discipline, coolness under fire, patient endurance in "many forced marches by day and night, under discomforts and "privations which all ranks have so cheerfully borne, during the "seven months the brigade served under his command in the "Khyber campaign.

"The 2nd Brigade, the Brigadier-General believes, has "borne its full share in the work allotted to the 1st Division, "and the perfect success which has crowned all the operations in "which it has been engaged, he attributes to the discipline, "cheerfulness and soldier-like qualities of the regiments under "his command.

"The 1–17th Regiment has been under the Brigadier-"General's command from the first operations to the present "time, and has shared with him the fatigues and privations of "the flank march to Kata Kushtia, both operations in the "Bazar Valley, and repeated expeditions, including the forced "march to Maidanak and the highly successful operation against "the Shinwarris at Deh Surak. The excellent conduct of the "regiment on all these occasions, as well as during the monotony "of less eventful camp life, has earned for it the lasting admira-"tion of the Brigadier-General, and, he believes, of the whole "Khyber army; and he begs to tender his best thanks to "Lieut.-Colonel Tompson, the officers, non-commissioned "officers and men of this gallant corps for their services while "under his command.

"The Brigadier-General looks on it as a compliment to the "brigade that they were selected for the honourable, though "very arduous, duty of rear guard in the final evacuation of "Afghanistan, it having been the first to enter and the last to "leave Afghan soil, and he had hoped to lead them to the last "moment back into Indian territory. It was only in obedience "to superior orders that he left them at Jellalabad to assume "another command.

"In bidding farewell to the brigade and warmly thanking "everyone connected with it, the Brigadier-General desires to "express the pride and satisfaction he has felt in being "honoured with the command of such a magnificent body of "soldiers, and the regret with which he now parts with the "brigade, a regret which is qualified by the reflection that two "of the regiments composing it will continue to serve under "him for some time longer in his new command."

The following is an extract from a report on the services of officers of the 1st Division, Peshawur Valley Field Force, by Lieut.-General Sir S. Browne, V.C., K.C.S.I. :—

"Her Majesty's 17th Regiment has been one of the most "useful in the Division. Its good discipline and the hearti-"ness with which it entered into any work it had to do, "reflects the greatest credit on Lieut.-Colonel Tompson, who "commanded it to my entire satisfaction, and on Captain "Anderson, the Adjutant, whose services are specially men-"tioned by his commanding officer."

Lieut.-General Earl Howe, C.B., was appointed Colonel of the regiment on the 25th June, vice General W. R. Faber, deceased.

The 1st Battalion formed part of the garrison at Lundi Kotal until the 14th August, during which time it suffered much from sickness (cholera, fevers and dysentery), Captain Gamble, 2nd Lieutenant Watson and 22 men having died during the two months' occupation of this place.

1st Battn. East Indies, Aug., 1879.

On the above date the regiment marched to Peshawur, arriving on the 16th, and had scarcely taken over the barracks, when cholera again made its appearance, in consequence of which the battalion marched *en route* for Nowshera, forming cholera camps at short marches, until the 2nd October, when it went into barracks there.

Her Majesty was graciously pleased to give orders for the appointment of Brigadier-General Cobbe and Lieut.-Colonel Tompson as Companions of the Order of the Bath.

1880

The 1st Battalion again proceeded to Peshawur, arriving on the 2nd May, where it remained until the 1st November, when the regiment commenced its march *en route* to Gwalior, Jhansi, and Nowgong, reaching its destinations by the 27th, with headquarters at Jhansi.

The system of the supply of army clothing by contract was continued up to this year, when, after a careful inquiry by a committee, it was decided that the clothing should in future be supplied from the Royal Army Clothing Department (with its factory at Pimlico), and that the allowance to commanding officers be withdrawn.

On the 16th October, Lieut.-Colonel C. G. Grant succeeded to the command of the 2nd Battalion, on Lieut.-Colonel Boyd retiring on half-pay.

1881

On the 6th February, the establishment of the 1st Battalion was increased by the arrival from England of a draft of two captains, two sergeants, three corporals and 200 privates.

On the 28th February, the annual inspection was made by Brigadier-General William Gordon, the inspecting officer intimating later that he had reported most favourably on the efficient state and conduct of the battalion.

In commemoration of the services rendered by the 1st Battalion during the campaign in Afghanistan, the regiment was permitted to bear on its colours the words "Ali Masjid" and "Afghanistan, 1878–79."—(London "Gazette," dated 17th June, 1881.)

On the 1st July this year, many important changes were made in the organisation of the army, and in the titles and uniform of regiments of Infantry of the Line and Militia.

The 27th Brigade (with all other infantry brigade depots) was abolished, and the regiment formed into a territorial regiment, as follows :—

Territorial Regiment.		Composition.	Head-quarters of Regmtl. Dist.	Uniform.		
Precedence.	Titles.			Colour	Facing	Lace
17th	The Leicestershire Regiment.	1st Battn. 17th Foot 2nd " " 3rd " Leicestershire Militia 4th " "	Leicester	scarlet	white	rose

All correspondence, returns, &c., were ordered to be addressed : " Leicestershire Regiment."

The establishment of officers of the regiment was also altered, and all officers with the rank of 2nd Lieutenant were to be styled Lieutenants, the former rank being abolished.

The following non-commissioned officers were also promoted to warrant grade, viz. : Sergeant-Major, Bandmaster, and Schoolmaster, if of 12 years' service, and re-engaged.

The re-engagement of private soldiers to complete 21 years' service was ordered to be discontinued, except by special authority from Army Headquarters, and then only granted in special cases.

The headquarters of the 2nd Battalion marched from Nusseerabad, *en route* to Jubbulpore, on the 1st December, and the detachment at Neemuch proceeded to Saugor at the same time.

Brigadier-General Carnegy, in a farewell order, expressed his regret at parting with a regiment " whose high state of " discipline, general good conduct and smartness had been " so conspicuous during the time of its being under his com- " mand, and for these results, alike creditable to Colonel " Grant and all ranks, he desired to offer his acknowledg- " ments, in the assurance that the battalion would, in the " future, uphold as 'The Leicestershire Regiment,' that good " name which it had always enjoyed as the 17th."

CHAPTER XIII.

HOME SERVICE, BERMUDA, BURMAH, ADEN, HOME SERVICE, NOVA SCOTIA, WEST INDIES, SOUTH AFRICA.

1882–1898.

1882

The headquarters of the 2nd Battalion arrived at Jubbulpore on the 20th January.

st
attn.
ngland,
382.

On the 31st January, the 1st Battalion was concentrated at Morar, where it encamped prior to proceeding to Bombay, *en route* to England, and, embarking on the 3rd and 4th February, the regiment sailed on the latter date for Portsmouth, where it arrived on the 9th March, and was quartered in the Clarence Barracks.

During the 11 years and 340 days the 1st Battalion was in India the following changes took place :—

Strength on arrival in India, 15th February, 1870 ..	799	Died of disease, killed, and died of wounds	209
Recruits joined from that period until the embarkation of the corps for England, 3rd February, 1882 ..	995	Discharged and invalided ..	909
Transfers from other corps ..	174	Transferred	326
Enlisted in India	7	Deserted	1
		Went out and returned with regiment : rank and file, 178 ; remainder who proceeded home with regiment, 352	530
Total	1,975	Total	1,975

On the 27th June, the 1st Battalion was fitted with the new valise equipment (introduced in 1871) and the old pack equipment was abolished.

On the 10th September, the battalion furnished a detachment of two officers and 80 men for duty at the magazine Marchwood, Southampton, which remained so quartered until the 14th March, following.

The Storming of Khelat. (The 17th Regiment taking the heights.)

(From an old print.)

On the 4th October, the battalion moved from the Clarence Barracks, Portsmouth, to the Anglesey Barracks, Portsea.

1883

On the 1st April, the establishment of the 1st Battalion was slightly augmented, showing a total of all ranks, as follows: 24 officers, two warrant officers, 39 sergeants, 40 corporals, 16 drummers and 480 privates; total, 601; and, on the same date, the ranks of Instructor of Musketry and Sergeant-Instructor of Musketry were abolished except in militia battalions.

On the 3rd October, the band of the 1st Battalion, proceeding to Leicester (on an invitation from the Mayor and Corporation), advantage was taken of the occasion to send the old regimental colours, which were accordingly escorted by two officers, the sergeant-major and four colour-sergeants, the whole regiment following as far as the railway station.

On the following day, the old colours (of a very large size) were deposited in St. Martin's Church, Leicester, over the monument to the memory of those who fell in the Crimea. The ceremony of laying them was conducted by the Rev. Canon Vaughan, vicar of the parish, and they were escorted to the church by the officers and men comprising the regimental district, the regimental band and colour party, with a guard of honour from the Leicester Volunteers.

1884

The 1st Battalion proceeded in two detachments (on the 19th March and 9th April) to Aldershot, where it was quartered in the Centre Infantry Block, Permanent Barracks, and joined the 2nd Brigade.

From the 1st April, the appointment of "Sergeant-Instructor of Musketry," which had been abolished, was again instituted, and the establishment of the 1st Battalion was augmented to 610 privates.

On the same date, khaki clothing, consisting of coat, vest, and trousers, was issued to the battalion, as an experiment, to replace the scarlet kersey and black cloth trousers, the scarlet tunic and black trousers being also issued. The khaki clothing, as an undress uniform, was to be reported on, after six months' wear, with a view to its general adoption throughout the Army, the scarlet being still worn as a full dress.

On the 14th July, Colonel W. D. Tompson, C.B., retired from the service, and the command of the 1st Battalion was assumed by Colonel A. H. Utterson, and, on the same date, Colonel Cecil M'Pherson obtained command of the 2nd Battalion on the retirement of Colonel Grant.

1885

The 2nd Battalion marched from Jubbulpore on the 7th February, to Lucknow, where it arrived on the 14th March.

On the 1st April, the khaki clothing, which had been issued experimentally to the 1st Battalion, was found to be unsuitable and condemned, the scarlet uniform being re-issued.

On the 4th December, new colours were presented to the 2nd Battalion at Lucknow by the Countess of Dufferin (Colonel C. M'Pherson being in command of the regiment), in celebration of which the officers gave a ball at the Chutter Munzil, which was very largely attended.

1886

On the 27th October, the 1st Battalion proceeded from Aldershot to York and was quartered in the Fulford Barracks.

The rank of 2nd Lieutenant was this year revived by the Royal Warrant of the 31st December.

In the observations on the Inspection Reports by the Commander-in-Chief, for the years 1885 and 1886, His Royal Highness the Duke of Cambridge remarked, each year, on his being perfectly satisfied with the condition of the 1st Battalion, which deserved much praise.

1887

In April this year, the 1st Battalion furnished detachments to Tynemouth, Leeds, Richmond, and Warrington, whilst the headquarters and three companies moved to a part of the 14th Regimental District barracks.

On the 20th June, the battalion took part in a review at York, in celebration of Her Majesty Queen Victoria's Jubilee on completion of fifty years of her reign, and on the 30th, the regiment was concentrated at Strensall, with the exception of the company at Tynemouth.

On the 1st July, Colonel A. H. Utterson retired on half-pay, and Lieut.-Colonel T. Braddell succeeded him in command of the 1st Battalion, and on the same date, Lieut.-Colonel S. Bradburne assumed command of the 2nd Battalion, on the retirement of Colonel C. M'Pherson.

On the 8th August, the headquarters and five companies, 1st Battalion, returned to York.

1888

On the 6th January, a fancy dress ball was given by the officers of the 1st Battalion at the Assembly Rooms, York, to celebrate the bi-centenary of the regiment. The officers were attired in the uniform of 1688, while those of the Leicester Militia wore the uniform of 1788, a guard of honour to receive the guests, being attired, half in the uniform of 1688, and half in that of 1888. All officers residing in the United Kingdom who had served in the regiment were invited to be present, and there was a large and brilliant assembly of guests. The officers of the battalion, in their fancy dress, were photographed in a group on the following day.

On the 5th April, a detachment of four companies, 1st Battalion, proceeded to Pontefract, until the 22nd May, and the half battalion from Strensall, and the company at Tynemouth, joined headquarters on the 23rd July.

1st Battn. Bermud 1888.

On the 6th September, the 1st Battalion embarked for Bermuda, and sailed on the following day, under command of Lieut.-Colonel T. Braddell, with a total of 18 officers and 800 rank and file, arriving on the 20th September. Shortly after the arrival of the headquarters and 500 men (five companies) at Prospect Camp, Bermuda, an epidemic of enteric fever broke out, resulting in the deaths of 22 men, whilst upwards of 130 men, women and children were attacked. The battalion, on arrival, had one company detached at Ireland Island (Naval Headquarters), and two companies at St. George's Island.

2nd Battn. Burmah 1888.

On the 13th November, the 2nd Battalion left Lucknow for Burmah, the headquarters, arriving at Thayetmyo on the

3rd December, detaching three companies to Myingyan and one to Salen in Upper Burmah.

In consequence of frequent raids made by rebels and dacoits amongst the Chin tribes, a detachment of 100 men, stationed at Myingyan, proceeded on the 28th December, under Captain Shaw, to join the Chin Field Force, Southern Division, and was employed in repelling raids, and in punitive expeditions against the Chins, until the following June.

1889

On the 16th January, a party under Lieutenant Glossop, whilst exploring a dacoit stronghold in the hills, came upon a body of about 100 men behind a strong stockade, whom they expelled, killing and wounding several, the firing being kept up at close quarters for about twenty minutes. On this occasion Lieutenant Glossop and the Burman magistrate were both severely wounded, also one private severely, who afterwards died, and one slightly.

Prior to leaving the Myingyan district, Brigadier-General Penn Symons issued a farewell order, specially mentioning Captain Bunbury, Lieutenant Alexander, Colour-Sergeant Brennan, and Sergeants Abbott and Worsdall, for their work in the field.

[2]nd Battn. [A]den, [1]889.

The 2nd Battalion left Thayetmyo, *en route* for Aden, on the 7th November, where it arrived on the 29th, having suffered a good deal from sickness during its eleven months' stay in Burmah.

1890

On the 5th February, Lieut.-Colonel W. M. Rolph succeeded Colonel T. Braddell in command of the 1st Battalion, and the arrival of a draft from England, on the 10th March, brought its establishment up to 907 non-commissioned officers and rank and file.

Lieut.-General J. C. Guise, V.C., C.B., was appointed Colonel of the regiment on the 13th June, vice General Earl Howe, C.B., transferred to the 2nd Life Guards.

[2]nd Battn. [E]ngland, [1]890.

On the 12th November, the 2nd Battalion embarked at Aden for England, arriving at Portsmouth on the 30th, and, on the 2nd December, proceeded to Warley.

On the 3rd December, Lieut.-Colonel C. F. W. Moir, from half-pay, assumed command of the 2nd Battalion on the retirement on half-pay of Colonel S. Bradburne.

1891

1st Battn. Nova Scotia, 1891.

On the 27th February, the 1st Battalion embarked at Bermuda in the troopship "Orontes," for Halifax, Nova Scotia, where it arrived on the 5th March, in a heavy snow-storm, after a long and stormy passage. Prior to its departure, Lieut.-Colonel Rolph was presented with the following address, signed by the leading inhabitants of Bermuda. Some libellous reports, which had no foundation, reflecting on the conduct of the regiment, had appeared in some of the American and Canadian newspapers, and, having been copied into the "Times" and other English newspapers, were entirely refuted by this address, which was published in the leading American and Canadian papers, and was well established by two letters from His Excellency, Lieut.-General E. Newdigate, to the "Times," more than endorsing the opinion of the people of Bermuda, as to the conduct and discipline of the battalion.

ADDRESS

By the Inhabitants of Hamilton to the Commanding Officer of the Leicestershire Regiment.

Bermuda, 24th February, 1891.

To Lieut.-Colonel Rolph,
Commanding 1st Battalion Leicestershire Regiment.

SIR,

We, the undersigned residents of Bermuda, desire, on the eve of the departure of the regiment under your command for Halifax, to take this opportunity of expressing our high appreciation of the exemplary conduct of the men during the term they have been stationed in these islands.

We have rarely met with a regiment composed of men who have conducted themselves in such a quiet and orderly manner, and, notwithstanding the fact of the barracks being in such close proximity to the town of Hamilton, we do not recall any instance in which the peace and quiet of the inhabitants have been disturbed by any misbehaviour of the men.

While, under ordinary circumstances, we should have taken pleasure in thus testifying to their general good conduct, we do so on this occasion with special satisfaction, as it affords us an opportunity of publicly entering our protest against the unwarrantable accusations, to the prejudice of the regiment, which have recently appeared in public prints abroad—accusations which, with our local knowledge of facts, leave us utterly at a loss to account for.

In bidding the regiment farewell, we would again refer with satisfaction to the cordial relations which have existed between it and the inhabitants of Bermuda generally, and, in expressing our regret that the battalion is so soon about to leave us, we feel assured that it carries with it the highest esteem and best wishes of the community at large.

(Signed by the Mayor and 65 Magistrates, Justices of the Peace, and other citizens.)

1892

The 2nd Battalion was armed with the magazine (Lee-Metford) rifle in November this year, and on the 19th December, the battalion proceeded from Warley to Chatham.

1893

st Battn. West Indies, 1893.

On the 25th March, the 1st Battalion sailed from Halifax, N.S., to the West Indies, with headquarters and four companies at Barbados, three companies at Jamaica, and one at St. Lucia.

1894

On the 5th February, Major C. W. Vulliamy was promoted Lieut.-Colonel, to command the 1st Battalion in succession to Colonel W. M. Rolph.

In July, the 2nd Battalion proceeded from Chatham to take part in the autumn manœuvres around Aldershot, and, in August, occupied the Malplaquet Barracks in the North Camp.

The Lee-Metford rifle was issued to the 1st Battalion, in the West Indies, on the 9th November. On the 3rd December, Major W. Gregg was promoted Lieut.-Colonel, 2nd Battalion, vice Colonel Moir, placed on half-pay.

1895

General Sir John Ross, G.C.B., was appointed Colonel of the regiment on the 6th February, vice Lieut.-General J. C. Guise, V.C., C.B., deceased.

On the 23rd December, the headquarter companies at Barbados (1st Battalion) embarked for Cape Town, and, from the 24th, its establishment was raised to 880 privates.

1st Battn. South Africa, 1895.

1896

The 1st Battalion (having left one company at St. Helena) landed at Cape Town on the 18th January, and proceeded to Wynberg, detaching one company to Simonstown, and was afterwards joined by the three companies from Jamaica.

On the 30th April, four companies, under Major Liardet, proceeded to Pietermaritzburg, Natal, to replace cavalry and mounted infantry ordered to Matabeleland, until the 28th September, when they rejoined headquarters.

On the 21st September, the 2nd Battalion left Aldershot to embark for Ireland, and arrived at Cork on the 22nd, where it took over barracks.

2nd Battn. Ireland, 1896.

1897

On the 10th February, Lieut.-Colonel C. W. Vulliamy was succeeded in command of the 1st Battalion by Major W. S. D. Liardet.

On the 18th July, 137 ponies arrived from Argentina for the 1st Battalion at the Cape, and were handed over to it to form a mounted infantry company.

On the 13th December, the 1st Battalion embarked for Natal, the company from St. Helena rejoining headquarters on board. On the 16th December, the battalion disembarked at Durban, and proceeded to Maritzburg, where it was encamped.

1898

On the 1st January, Major F. W. Reader was gazetted Lieut.-Colonel to command the 1st Battalion, vice Lieut.-Colonel Liardet retired.

On the 15th June, one company of the 1st Battalion and a detachment of mounted infantry (in all four officers and 150 men) proceeded to Eshowe, Zululand, and on the 15th August, 16 officers and 511 men proceeded on manœuvres to Ladysmith, by march route (marches 12, 7, 14, 16, 18, 20 and 20 miles), carrying complete service marching order kit and blank ammunition—weight, exclusive of clothes, 39 lbs. 13 ounces. They returned by rail on the 3rd September.

On the 15th August, the 2nd Battalion left Cork to take part in the manœuvres on Salisbury Plain, and, on return to Ireland, occupied quarters, on the 9th September, at the Curragh Camp.

On the 10th November, Lieut.-Colonel Reader died, and, on the 3rd December, Major G. D. Carleton was promoted to command the 1st Battalion.

On the 3rd December, Lieut.-Colonel Gregg relinquished command of the 2nd Battalion, on expiration of four years in command, and on the 7th, Major A. W. M'Kinstry was promoted Lieut.-Colonel, vice Lieut.-Colonel Reader, deceased, and posted to the 2nd Battalion.

Officer's Shako Plate (*reduced size*).
1843–45.

Officer's Breastplate (*exact size*).
1843–45.

CHAPTER XIV.

THE SOUTH AFRICAN WAR, EGYPT, EAST INDIES, AND HOME SERVICE.

1899–1910.

1899

ON the 9th January, the mounted infantry company, 1st Battalion, under Captain Sherer, marched to Nottingham Road to encamp, in order to avoid horse sickness.

On the 25th January, the arrival of a draft from England brought the strength of the 1st Battalion up to 1,026, exclusive of officers.

On the 18th March, the headquarters 1st Battalion proceeded by rail to Mooi River to encamp, pending their move to Ladysmith, which was effected in three marches, commencing 2nd May, the mounted infantry having previously rejoined, and the detachment from Eshowe rejoined on the 27th May.

At midnight on the 24th September, the 1st Battalion proceeded by rail to Glencoe *en route* to Dundee, and the following officers entrained with it :—

Lieut.-Colonel Carleton (commanding).
Major E. R. Scott.
„ R. L. Sandwith.
Captain F. E. Glossop.
„ H. L. Croker.
„ C. G. Blackader.
„ T. M. Drew.
Lieut. W. Bryce.
„ R. N. Knatchbull.
Lieut. E. L. Challenor.
„ C. H. Haig.
„ F. Lewis.
„ D. Faichnie.
„ C. P. Russell.
2nd Lieut. T. N. Puckle.
„ S. O. Everitt.
Captain and Adjutant H. M. Welstead.
Lieut. and Quartermaster W. Baker.

On arrival at Glencoe at 6 a.m. 25th, the battalion marched to the bivouac ground, five miles from the station.

On the 27th, the battalion was encamped about two miles from Dundee, to await developments, with a mixed force, consisting of the 18th Hussars, three batteries R.F.A., 1st King's Royal Rifles, 2nd Royal Irish Fusiliers, and the mounted infantry of the Royal Dublin Fusiliers and of the battalion, the latter being commanded by Captain L. C. Sherer with Lieutenants Weldon, Paul and Hannah.

The South African War.

War was declared on the 12th October, 1899, by the South African Republic and by the Orange Free State against Great Britain.

It was arranged for the four infantry battalions to take in turn the outposts of the force, and the brigade stood to arms daily at 4.30 a.m.

On the 20th October, after having been dismissed, the Dutch opened fire at about 5.45 a.m. with two quick-firing guns from Talana, on the camp, at a range of some 4,000 yards, the first casualty being a trumpeter of one of the field batteries. The battalion was directed to hold the right flank and part of the rear of the camp, together with the 67th Battery, as it was known that the enemy had possession of the Newcastle Railway, north of Glencoe Station, whilst the mounted infantry company (under Captain Sherer) was sent to watch the extreme right of the position taken up by General Sir Penn Symons.

During the attack on Talana Hill the General fell mortally wounded.

On the 21st October, the brigade stood to arms at 4.30 a.m., and moved out of camp to a ridge north-east of the Dundee and Glencoe Railway. The enemy's guns opened fire about 4 p.m. at a range of 8,000 to 10,000 yards and quite outranged our field batteries. The Dutch guns were placed on a low spur of the Impati Mountain, and kept up a fire until dusk. Lieutenant M. Hannah, of the battalion, was killed by the base of a shell, and one private of the mounted infantry company was wounded. The battalion, together with the 1st Royal Irish Fusiliers and 1st King's Royal Rifles, bivouacked for the night on a small kopje, in rear of the position and out of range of the enemy's guns. With a mixed force, on the

22nd October, it then commenced its march to Ladysmith, which was reached on the 26th after some trying marches and in torrents of rain throughout, but the route was not interrupted in any way by the Dutch.

On arrival at Ladysmith the battalion occupied quarters at " Tin Town " (so called by the Boers on account of its iron huts), and remained at Ladysmith throughout the siege whilst the place was entirely isolated. The line of rail at this period was loaded with rations for the troops (men and animals), but three months' supply for 5,000 men had fallen into the hands of the enemy, also a large quantity of ammunition, tents, and other equipment.

On the 29th October, the battalion formed part of the 8th Brigade, when it became engaged in the action at Lombard's Kop, on which occasion, the four mules of its Maxim gun having been shot and the native drivers having run away, it became subjected to a very heavy fire from all kinds of projectiles and the gun had to be abandoned. Corporals Gillespie and Harris, for their endeavour to recover it by hand, were later awarded the Distinguished Conduct Medal for service in the field. The casualties were three men killed and 18 wounded.

On the 2nd November, the battalion marched out of Ladysmith and took up the following positions : Five companies at Cove Redoubt, two companies at Gordon Post and one company at the railway station in charge of the armoured train.

During its stay at Ladysmith the battalion was constantly employed in sorties and reconnaissances, and from the 2nd November, 1899, to 13th March, 1900, it did not cease to bivouac all through the wet season.

On the 7th November, three men of the regiment were severely wounded by a Long Tom shell, one of whom died the same day.

On the 9th November, the naval guns fired a Royal Salute of 21 guns with ball ammunition, in the direction of the Boer headquarters' laager, in honour of the Prince of Wales's birthday.

1900

On the 6th January, when the assault on Cæsar's Camp by the Dutch was proceeding, a small attack was made on

Observation Hill West, but, on the leader and a few of the enemy being killed, they withdrew and did not endeavour to attack the position again.

nd attn. gypt, 900.

On the 7th February, the 2nd Battalion embarked at Queenstown for Egypt, under Lieut.-Colonel A. W. M'Kinstry, arriving at Alexandria on the 18th February.

All through the months of January and February, 1900, the 1st Battalion had a large amount of sickness, with the rest of the Ladysmith garrison, as many as 400 being in hospital, and Lieutenant Russell died from enteric fever and some 50 non-commissioned officers and men. Rations were greatly reduced during these months. The first issue of chevril soup (from horseflesh) was on the 1st February, but for several weeks before, the daily ration had been horse and muleflesh, the latter being generally considered the better. The scale of ration on the 8th February was ½ lb. of meal, one ounce of sugar, and 2 lbs. of horseflesh. On the 10th February, it was reduced to ¼ lb. of meal, 2 lbs. of horseflesh and one pint of chevril per man.

Ladysmith was relieved at 6.25 p.m., on the 28th February, after a siege of 121 days.

The following officers of the 1st Battalion served throughout the siege :—

Lieut.-Colonel G. D. Carleton, commanding.

Major E. R. Scott.
Captain L. C. Sherer, commanding Mounted Infantry Company.
Captain H. M. Welstead (Adjt. to 26th Nov., 1899).
" F. E. Glossop.
" H. L. Croker (Adjt. from 26th Nov., 1899).
" C. G. Blackader.
" T. M. Drew.
" C. E. Cox.
Lieut. W. Bryce (promoted Capt., 19th Jan., 1900).

Lieut. R. N. Knatchbull.
" B. C. Dent, Mtd. Inf. Cy.
" E. L. Challenor.
" J. R. Paul, Mtd. Inf. Cy.
" C. H. Haig.
" D. Faichnie.
" W. H. Young.
" F. Lewis.
" H. M. Travers.
2nd Lieut. T. N. Puckle.
" S. O. Everitt.

On the 1st March, four companies, under Major E. R. Scott, attacked Bell's Kopje and laager, and found several hundred rifles, a large quantity of ammunition, chiefly Mauser, but many soft-nosed and split. One prisoner was taken, and no casualties on our side.

On the 2nd March, the following special Natal Field Force Order was published by Lieut.-General Sir George White, V.C., G.C.B., commanding at Ladysmith :—

" Now that the relief of Ladysmith garrison has been so " gallantly accomplished by the force under the immediate " command of General Sir Redvers Buller, V.C., K.C.B., who " has assumed command at Ladysmith, Lieut.-General Sir " George White wishes to thank the officers and men of the " Royal Navy, and of the Imperial and Colonial forces of the " Ladysmith garrison, for their soldier-like conduct throughout " the trying siege. It will be a matter of the highest satis- " faction to every man in the garrison that his efforts and " endurance have played no small part in holding Ladysmith " (which was the first and most coveted objective of the " united armies of the Free State and the South African " Republic) for our Queen and country, and that his successful " resistance of the enemy has been acknowledged in terms of " the highest sympathy and admiration by Her Majesty and " by the country at large."

On the 3rd March, the battalion took part in lining the road between the town and Tin Town to receive Sir Redvers Buller and his army. This parade showed the weak state of the defenders, many of whom constantly fell out from weakness.

On the relief of Ladysmith the following message was received from the Queen :—

" Thank God that you and all those with you are safe, " after your long and trying siege, borne with such heroism. " Congratulate you and all under you from the bottom of my " heart. Trust you are not all very much exhausted."

Congratulations were also received from the Commander-in-Chief, from the Governors and Premiers of the different colonies, from the Viceroy, and Commander-in-Chief in India, from the Government and Members of Parliament, and from most of the great towns and societies in the United Kingdom.

On the 12th March, the newly constructed 8th Brigade (which now consisted of the 1st Liverpool, 1st Leicestershire, 1st and 2nd Battalions King's Royal Rifles) marched to Ouderbrook *en route* to Colenso, where the brigade was sent to recruit its health, under Major-General F. Howard, C.B., A.D.C.

On the 16th March, Lieutenant F. D. Harrison and 150 reservists joined, and on the 24th, the Volunteer Service Company joined the battalion at Colenso, under command of Captain W. A. Harrison with Lieutenants H. N. Rowlatt and C. H. Jones.

On the 2nd April, the 8th Brigade, 4th Division, marched from Colenso to Ladysmith, and the battalion, on the 10th, to Gun Hill and Lombard's Kop, where it was employed in strengthening the positions of the Kop and neighbouring hills, the men working continually ten to twelve hours a day for some weeks.

On the 21st April, 250 reservists joined the battalion, mostly Section D, Army Reserve.

On the 10th May, the battalion, under Lieut.-Colonel Carleton, together with a squadron 18th Hussars and a battery R.F.A., marched to Pieter's Farm (to the east of Ladysmith), where it was detached to hold the communications on account of General Buller's advance to Dundee. On the 15th May, the march was continued to Modders Spruit, where they joined the rest of the 8th Brigade, and, on the 16th, the force marched to Sunday's River, proceeding, on the 24th, to Waterkloof Farm in the Biggars Berg, on the 25th to Kalebus Farm, and, on the 26th, to Ingagane, where the railway station and bridges had been wrecked by the Dutch.

On the 28th May, the following force, under command of Major-General the Honourable N. G. Lyttelton, marched to Cattle Drift on the Buffalo River, viz. :—

The 18th Hussars : Three batteries R.F.A., one howitzer battery, two 12-pr. naval guns ; one pom-pom ; and the 8th Brigade, which consisted of : The 1st Battalion King's Liverpool Regiment, 1st Battalion Leicestershire Regiment, 1st Battalion King's Royal Rifles, 1st Battalion Gordon Highlanders and the 4th Division Mounted Infantry Battalion of four companies, including that of the battalion under Captain L. C. Sherer with Lieutenants Dent and Beamish.

On the 29th May, a battery R.F.A. and the 1st Liverpool and 1st Leicestershire Regiments were detached under Lieut.-Colonel Carleton, to watch and command Steil's Drift, over the

Buffalo River, and on the following day moving across the river, they joined General Lyttelton's force at Inkuba Spruit. On the 1st June, the force marched to Wool's Drift; on the 2nd to Coetzee's Drift and encamped until the 16th.

On the 4th June, the Mounted Infantry Battalion and some of the 18th Hussars endeavoured to get into Wakkerstrom but failed, the enemy being found in strength in the neighbouring hills.

On the 16th June, the battalion formed part of a small force, under command of Major-General Howard, in which were included the 1st King's Royal Rifles, Mounted Infantry Battalion, and a battery R.F.A. On marching to Slang's Spruit the Mounted Infantry Battalion entered Wakkerstrom without any casualties to our force. On the 19th, the force returned to Ingogo Heights, where the battalion was put into tents.

On the 29th June, a draft of militia reservists joined the regiment, which remained at Ingogo until the 3rd July, when it marched with a force under command of Major-General Brocklehurst (commanding 1st Cavalry Brigade Natal Army) to Laing's Nek, and on the following day to Zand Spruit. On the 17th the enemy shelled our camp from Grass Kop.

On the 21st July, four companies of the regiment, under Captain Melvill, proceeded to Meerzicht, and, on the 24th, assisted in the capture of the Rooi Kopjes, a position south of Amersfort, that had been held with great doggedness by the Boers. Grass Kop had also been taken on the 22nd July. The remainder of the battalion marched to Meerzicht on the 29th July, and joined General Buller's force, which concentrated here before advancing to Amersfort.

On the 6th August, the following force marched from Meerzicht: 1st Cavalry Brigade, under Major-General Brocklehurst; 2nd Cavalry Brigade, under Lord Dundonald; 4th Division, under Major-General the Honourable N. G. Lyttelton, comprising the 7th and 8th Brigades; a battery R.H.A.; three batteries R.F.A.; a howitzer battery; two 4·7 naval guns; four 12-pounder naval guns, and two pom-poms.

On the 7th August, Amersfort was taken, the battalion being on the right of the line, and its casualties were three men wounded; the Maxim gun was also slightly damaged by a splinter of a shell. The whole force bivouacked there for the night, and, as the baggage waggons did not arrive, the troops were without blankets, the thermometer being as low as 27° Fahrenheit.

On the 9th August, the march was resumed to Ermelo, which was entered on the 12th. On the 16th, the troops proceeded to Twyfeler, and, remaining stationary, entrenched all the neighbouring positions, the battalion having a portion of the defences allotted to it.

On the 21st August, the force marched to Van Wyk's Vlei, and came in contact with the enemy, who eventually held the Bergendal position with strength and tenacity. On the 23rd, the march was continued to Geluk Farm, where the Boers sniped the camp all day and far into the night, the casualties of the battalion being four privates wounded. On the 24th, there was a slight earthquake.

On the 27th August, General Buller began his attack on the Bergendal position,* which was held in great force by the enemy, and was looked upon as their final stand before retreating into the mountainous regions around Lydenburg and Origstadt. The position was a very long one, but the key was the kopje near the farm of Bergendal. General Buller commenced an artillery duel at about 6.30 a.m., and practically silenced the Boer guns early in the day. At about 2.30 p.m., the 2nd Battalion Rifle Brigade, supported by the Gordon Highlanders, and the 1st Royal Inniskilling Fusiliers, supported by the battalion, attacked the kopje with a very wide front. The position was taken at about 4.30, and the enemy let a pom-pom (Vickers-Maxim) fall into our hands. The team was destroyed, and the Boers fought the gun until the assailants were within 200 yards.

On the 28th August, the force marched to Machadadorp, on the railway, and on the 30th, encamped at Helvetia.

On the 1st September, the march was continued to Lydenburg, and on the 2nd, when encamped in the Badfontein Valley, on the Crocodile River, the battalion was attached to the 7th Brigade, under Major-General Walter Kitchener. On the 4th September, the enemy shelled the brigade with a

* Known as "Belfast."

6-inch Creusot gun (a Long Tom), necessitating a move back out of range. On the 5th, the battalion, with the 1st King's Royal Rifles, attacked and took the high ground and hills on the south-east of the Badfontein Valley, but as the firing line gained the summit of the hill the enemy had removed the gun and were in fast retreat to Lydenburg. The battalion gained a mention in General Buller's despatches for this work.

Shortly after having bivouacked, on arrival at Lydenburg (8th September), the enemy fired on the force with two Long Toms, and also into the town, when the battalion had five men wounded. The regiment now formed working parties for strengthening the defences of Lydenburg, and also had one company detached to form a strong post with night pickets, at a farm at the east of the town, whilst two companies took convoy duty in turn with two companies of the 2nd Battalion Rifle Brigade between Lydenburg and Badfontein.

On the 30th September, five companies of the battalion joined a small force and marched to Kruger's Post, the movement being in connection with General Buller's advance from the Devil's Knuckles. After driving a few of the enemy off the ridge, a few miles from the post, the battalion bivouacked, the guns being brought up into position, by which General Buller was materially helped. During the day (1st October) the two forces met, and at about 7 p.m. the enemy shelled the camp, killing one officer and nine men of the Devonshire Regiment and one officer and 10 men of the South African Light Horse. On October 3rd, the battalion returned to Lydenburg.

Lieut.-Colonel Carleton having now assumed command of the 8th Brigade (which consisted of the 1st Royal Inniskilling Fusiliers, 1st King's Royal Rifles and 1st Gordon Highlanders), the headquarters of the battalion, under Captain Welstead, left Lydenburg (October 9th) with the 18th Hussars and some guns, under command of Lieut.-General the Hon. H. Lyttelton, for Middelburg, where it arrived, after meeting with very little opposition, on the 16th October.

The battalion now took part in strengthening the defences of Middelburg, having one company detached at Great Oliphant's Railway Station, one at Little Oliphant's River, one at Rockdale Farm (about four miles off), and the remainder in the vicinity of Middelburg. Captain Welstead was in command of the battalion from October 9th to December 7th, when Captain Sherer resumed command, on rejoining from the 4th Division Mounted Infantry.

1901

Lieut.-Colonel Carleton having been appointed, on the 7th January, Commandant of Middelburg (Transvaal) on the breaking up of the 8th Brigade, Captain Sherer continued in command of the battalion until the 29th April.

The battalion now took part in two or three small expeditions, but did not succeed in doing much more than burning a few farms and driving the enemy back a few score of miles, the latter returning, on the return of the column, to Middelburg.

On February 1st, the battalion formed part of a column under Lieut.-Colonel W. P. Campbell, 1st King's Royal Rifles, consisting of the 18th Hussars, one company Mounted Infantry, two guns and one naval 12-pounder. The column was one of seven that swept the country between the Natal railway and the Delagoa line to the borders of Swaziland, and was under the supreme command of Lieut.-General Sir John French. Its route being towards the other columns, it came in contact with that of General Smith-Dorrien on the 7th February, the day after he had been attacked at Lake Chrissie. After getting in touch with the other columns, the advance was pushed forward with great vigour, the route that Colonel Campbell took having been via Hamilton, Newstead and Amsterdam, that part of the Transvaal called the New Glasgow country. Near Roodeport the casualties of the battalion were five men wounded.

Amsterdam was reached on the 14th February, and Piet-Retief on the 28th, where several of the columns met. There is little to relate here, except that a large number of Boer families were collected, who were sent into the Concentration Camp, and their farms and foodstuffs destroyed. Piet-Retief was made into a depot, and on the 1st March, Colonel Campbell's column was pushed forward to the Assegai River. Here the battalion put the position above Sand Drift into a state of defence, and was one of the blocks in connection with the other columns to prevent the enemy breaking back from the advance of the mounted troops along the Swaziland border.

On the 6th March, the battalion moved to Unchange Drift, on the Assegai River, Swaziland, and found the Swazis friendly in giving every assistance and information regarding the enemy, this being the only column that had up to date been in Swaziland. The weather was very bad, with continuous rain, since the 10th February, the only covering the men had being one blanket and one waterproof sheet, no tents being carried.

On the 9th March, the battalion returned to camp near Sand Drift on the Assegai River, and remained entrenched until the 10th April, when, with the remainder of Colonel Campbell's column, it marched to Piet-Retief, which was evacuated on the 14th, the column pursuing its course to Middelburg, which was reached on the 27th April. On the 28th, Major Griffith joined the battalion for duty, and on the 29th, Major Scott resumed command on return from sick leave. On Major Sherer relinquishing command he was appointed Assistant-Provost Marshal to Colonel Campbell's column, Captain Melvill remaining with the column as Intelligence Officer.

On the battalion arriving at Middelburg it took over part of the defences there, comprising: the southern part; the railway station and defences; Rockdale Farm, Great Oliphants, Little Oliphants (the two latter being on the railway east and west of Middelburg), Witbank Station and also Witkyk, another new post on the railway. The battalion garrisoned the above posts until the 25th July, when it proceeded in brigade under Colonel Campbell, whose column now consisted of the 18th Hussars, 300 West Australians, four field guns and the battalion, the whole under the supreme command of Major-General Walter Kitchener, whose column was also working in the same part of the country, viz., north of Pretoria and Delagoa Bay line.

From the 8th to 25th August, the battalion guarded a supply camp some thirty miles from the line (Rooikraal), and on the 25th moved with the two columns towards Wonderfontein, which was reached on the 4th September. About the 27th August, when a small empty convoy under command of Captain Welstead was returning to Middelburg, Second Lieutenant Tronson was slightly wounded, and one man severely, who afterwards died.

At this period the principal work for the infantry was guarding food depots and helping to escort convoys, whilst the mounted troops of both columns were having very hard and difficult work in pursuit of the Boer general, Ben Viljoen, who on more than one occasion turned the tables and ambushed some of his pursuers.

On the 9th September, the mounted troops of both columns started south, and the battalion, with a few mounted men and sick horses and all the ox transport (under command of Major E. R. Scott) proceeded in rear of the mounted troops to Ermelo, which was reached on the 12th September, and burnt to the ground. Two days later Amersfort shared the same fate.

From Amersfort the battalion marched to Zandspruit Railway Station and thence to Utrecht (26th September) and Vryheid, and conducted two convoys to Botha's Farm, near which General Kitchener engaged the enemy, who held the ridge above the farm with great determination, only leaving under cover of darkness.

From this district the battalion marched to Paulpietersdorp, and placed this town in a state of defence, remaining there from the 7th to the 18th October, and thence marching to Luxemburg, took part in the searching of the Kloof with the combined columns of General Walter Kitchener, Colonels Campbell, Pulteney, Stuart and General Hamilton, which resulted in a very large seizure of cattle and in families being taken to the refuge camps, and also in taking some two to three hundred prisoners.

The battalion now came to the railway line with the remainder of the columns, and reached Volksrust on the 27th October, having marched eighty-four miles in four days. From Volksrust the battalion was railed to Standerton (November 5th), where tents were issued and the whole regiment encamped, and obtained new clothing, new boots and other equipment.

Here arrangements were made for building the new Block House Line, known as the Standerton-Ermelo Block House Line, and on the 13th November, Lieut.-Colonel Carleton, having given up the command of Middelburg, assumed command of the battalion.

The Block House Line was built thus :—

Every ten miles there was a strong fortified post with a garrison, roughly of 100 men. The block houses were 800 yards apart, and were occupied by a non-commissioned officer and six men.

Around each house was a barbed wire entanglement, and between each a barbed wire fence of six strands of wire and a curtain each side of five strands. During the time the battalion occupied this line a ditch was dug on either side of the fence five feet wide at top, four feet deep and three feet wide at bottom.

The occupation of these block houses proved very tedious and trying for both officers and men, for, as each only contained one non-commissioned officer and six men, the look-out had to be continuous and the sniping, which usually took place at night, kept all garrisons awake.

Also many small intermediate works had to be built and garrisoned, in order to prevent crossings of the wire, and although the battalion guarded nearly 35 miles of block house line, there was never a serious crossing effected, the total being only thirteen, and these were principally made by single men carrying despatches. Every crossing was known as the wire had to be cut, and each wire of the fence had an alarm at both ends ; also batteries of rifles were fixed so as to fire down the wire, six rifles being fastened together, carefully laid and aimed before sunset.

On the 28th November, a draft of 55 non-commissioned officers and men joined from England, most of whom had been invalided from South Africa during the war.

The part of the block house line which the battalion first held was from Standerton to Blesbok Spruit, 32 miles, and it was completed by the 5th December.

1902

On the 12th February, the 1st Battalion was further augmented by an increase of 120 men from the 2nd Battalion in Egypt.

About the 2nd March, the Post of Morganzone having been handed over to the battalion, caused a further extension of another five miles of block house line and another strong post.

The posts held by the battalion consisted of :—Leeuwspruit post : headquarters of battalion and two companies ; Witkyk post : headquarters of three companies ; Blesbok Spruit : headquarters of one company ; Morganzone post : headquarters of one company ; whilst one company held a small post in the centre of the line five miles from Standerton.

The battalion occupied the block house line until peace was declared in accordance with the Treaty of Vereeniging, when the line was dismantled by it, to the extent of collecting all wire and standards, leaving 2,000 yards of wire at every block house, and the transport for the whole took 178 ox waggons to bring into Standerton, to which station the battalion moved in July, and thence to Platrand until the 8th September, when it marched to Volksrust, arriving on the 10th.

Whilst at these stations nearly 500 reservists were sent home, and, on the battalion receiving orders to proceed to India, 250 non-commissioned officers and men were left in South Africa, to go home, who had completed six years' service with the colours.

The battalion was encamped at Volksrust until it moved to Durban on the 6th November in two trains of open trucks, to embark for India.

st Battn. ;ast ndies, Iov., 902.

The regiment sailed on the 7th for Madras, and, arriving on the 30th, disembarked on the 1st December and proceeded to Fort George, detaching a half battalion to Bellary under Captain W. S. Melvill.

Lieut.-Colonel Carleton retired from the service on December 4th, and handed over the command to Brevet-Major and Adjutant H. L. Croker.

The strength of the 1st Battalion on landing in India was 13 officers and 458 non-commissioned officers and men.

On the arrival of the 1st Battalion at Madras, on the 1st December, the following address, signed by the chairman of the Municipal Council, was handed to Lieut.-Colonel Carleton, together with a purse containing Rs. 1000, which was expended in Christmas Dinners:—

"The public of this city has deputed me in its name to "welcome you and all ranks of your distinguished regiment "on your arrival in Madras. Local conditions unfortunately "prevent such a public demonstration as would visibly testify "our gratitude and admiration for the gallant services the "'Leicestershires' have rendered in their last campaign.

"But our sentiments are none the less permanent or "sincere that we are driven to compress, and voice them in "the formal compass of a letter. India, not less than England, "is unlikely to forget the steadfast courage with which your "regiment met and repelled the first wave of the Boer attack "in Natal, and all who knew anything of the history of your "corps, knew that in your splendid isolation at Ladysmith "you would be true to the regimental traditions of two hundred "years, and emerge ultimately as triumphant in South "Africa as you emerged in Flanders and Holland and Germany, "at the taking of the Canadian Louisburg, at the capture of "Ghuznee and Khelat, from the sufferings of the Crimea, or "at the wresting of Ali Masjid from your Afghan foe.

"We can add nothing to the last long roll of your "latest-won distinctions, whose catholicity best testifies to "the spirit which animated all ranks of your regiment from "colonel to private. Yet, in the hope that you yourself, and "all under your command, may be disposed to attach some "friendly value to this imperfect public appreciation by your "fellow citizens, of all castes and creeds, of duty well and "faithfully discharged, before a world watching in breathless "suspense, I tender, on their behalf, to your non-commissioned "officers and men, a purse filled by public contribution for "the celebration of the approaching Christmas festivities, and "greeting to yourself, your officers, and every member of your "regiment who, by the legend of Hindoostan upon your colours, "have long ago been presented with the freedom of India."

To this Lieut.-Colonel Carleton replied by letter, dated Madras, 2nd December, as follows:—

"I have the honour to acknowledge the receipt of the "complimentary address and welcome of the Municipal "Committee of the City of Madras to the battalion under my "command on its arrival in India, and to thank them, on the "part of the officers, non-commissioned officers and men, for the "good wishes expressed, and for the very handsome donation "accompanying the welcoming letter. This will be very much "appreciated, at the next Christmas festivities, by the men.

"It gives us great satisfaction to know that the services "of the battalion in the past have met with the approval of "the country, and I trust that in the future it will be afforded "further opportunities to add to the reputation it already "bears, after long service in the Indian Empire."

Throughout the South African War the total numbers killed in action, and died from wounds and disease, were three officers (Captain H. C. Thorald and Lieutenants W. M. Hannah and C. P. Russell) and 122 non-commissioned officers and men.

The rewards to officers and men were as follows: Captains C. G. Blackader and Lieutenants R. N. Knatchbull and J. R. H. Paul were made Companions of the Distinguished Service Order, and Captains H. M. Welstead, F. E. Glossop, H. L. Croker and A. H. Wilkinson received the brevet rank of major. All the above officers were also twice mentioned in despatches, with the exception of Captain Wilkinson and Lieutenant Knatchbull mentioned once, and Captain Croker and Lieutenant Paul mentioned three times.

The following officers were also mentioned in despatches: Major L. C. Sherer, Captains G. I. Walsh and W. Bryce, Lieutenants B. C. Dent (twice), B. Weldon, B. C. Dwyer, C. H. Haig (twice), T. N. Puckle (twice), and D. Y. Watt.

The Distinguished Conduct Medal was awarded to Colour-Sergeants A. Wood and Jones, Sergeant Smart, Corporals M. Gillespie, C. Harris, and J. Withers, Lance-Corporal J. Bradshaw, and Privates R. Allen and F. Green.

The following medals were given: The Queen's South African for the year 1900, which carried with it the following clasps for actions in which the battalion was engaged, viz., Talana, Defence of Ladysmith, Laing's Nek and Belfast (2), the King's Medal, 1901 and 1902.

The honours to be borne on the colours were: "South Africa, 1899–1902" and "Defence of Ladysmith."

On the 11th November, Major E. R. Scott was promoted Lieut.-Colonel to command the 2nd Battalion in succession to

Colonel A. W. M'Kinstry, placed on half-pay on expiration of command.

On the 30th November, the 2nd Battalion embarked at Alexandria for conveyance to Southampton, and was transhipped for passage to Guernsey and Alderney, arriving on the 15th December.

2nd Battn. England, Nov., 1902.

1903

On the 24th February, Lieut.-Colonel J. G. L. Burnett assumed command of the 1st Battalion on promotion from the 2nd.

Major-General G. T. Brice was appointed Colonel of the regiment on the 29th July, vice General Sir John Ross, G.C.B., deceased.

On November 12th, at 3.30 p.m., a special service was held at St. Martin's Church, Leicester, conducted by the Vicar, the Rev. Canon Sanders, M.A., LL.D., for the purpose of unveiling a memorial window and tablets, erected to the memory of the officers, non-commissioned officers and men of the four battalions of the regiment (including Militia and Volunteers) who were killed in action or died of wounds and disease throughout the South African War, 1899–1902. The band of the 2nd Battalion attended, and, in the course of a most impressive service, a short address was delivered by Major-General Utterson, C.B., who then proceeded to unveil the window and tablets.

Some most important changes took place this year in the administration of the Army.

On the 7th November, the War Office appointed a Reconstitution Committee to assemble for the re-organisation of the Army, presided over by Viscount Esher, with Admiral Sir James Fisher and Colonel Sir G. S. Clarke, members, when it was ordained that :

The post of Commander-in-Chief should be abolished, and in lieu there should be appointed a council of senior general officers, specially selected, to be termed "The Army Council," in conjunction with the Secretary of State for War, and with the abolition of the Commander-in-Chief, a new office, viz., that of Inspector-General of the Forces was created.

With a view to the introduction of entirely new ideas, officers holding the principal staff appointments, such as Director-General of Mobilization, Adjutant-General and Quartermaster-General, &c., at the War Office, were directed to vacate them.

1904

Field-Marshal the Earl Roberts accordingly ceased to be Commander-in-Chief from the 11th February, the Army Council being installed from the following day.

The Finance Department of the army was also entirely re-constructed on a civil footing, with a view to abolishing the Army Pay Department, which, however, was re-installed in 1909.

The Committee, moreover, were in favour of the restoration of the old regimental numbers to battalions as their first titles, when unlinked for purposes of the mutual supply of drafts, by which means "*the connection between the regiments and the*
"*finest pages in the history of the British Army would be re-*
"*established, and at the same time the great convenience of the*
"*numbers—in war time especially—would be regained.*"

This recommendation, however, was not acted on, though the regimental numbers have since been retained in the Army List.

On the 9th September, Lieut.-Colonel J. G. L. Burnett, commanding 1st Battalion, died at sea when proceeding home from India, and, from the 10th, Major G. H. P. Burne was promoted Lieut.-Colonel and appointed to the command.

On the 28th September, the 2nd Battalion left Guernsey and Alderney for Colchester, where it arrived on the following day.

1905

On the 16th January, the half of the 1st Battalion at Bellary marched to Belgaum, arriving 9th February, and was joined there by the headquarters from Madras on the 31st March.

On the King's birthday this year, in celebration of the Jubilee of the Crimean War, Major-General G. T. Brice (Colonel of the regiment) was nominated a Companion of the Bath.

On the 23rd September, Major-General A. H. Utterson, C.B., was appointed Colonel of the regiment, vice Major-General G. T. Brice, C.B., deceased.

1906

The 2nd Battalion embarked at Southampton on the 21st September for Bombay, and proceeded to Belgaum, in relief of the 1st Battalion, arriving 13th October.

2nd Battn. East Indies, 1906.

On the 9th October, the 1st Battalion was armed with the new short rifle, M.L.E. and equipment, which had been issued to the 2nd Battalion in England, in April.

The following is an extract from Brigade Orders, dated Belgaum, 15th October :—

" The 1st Battalion Leicestershire Regiment being about " to leave the brigade, the Colonel on the Staff wishes to place " on record his great appreciation of their soldierly qualities, " both in camp and quarters, and of their uniform good " conduct. Their successes at all manly sports have been " a source of pride and great pleasure to him."

The two battalions were together at Belgaum until the 16th October, when the 1st Battalion (strength, 13 officers and 466 rank and file) left for Bombay, *en route* to England, where the regiment arrived on the 9th November, and proceeded to Shorncliffe. Here it was joined by details of five officers and 319 non-commissioned officers and men, who were awaiting its arrival.

1st Battn. England 1906.

On the 11th November, Major V. T. Bunbury, D.S.O., was promoted Lieut.-Colonel to command the 2nd Battalion, vice Colonel E. R. Scott, retired on half pay.

1907–08

The 2nd Battalion was adjudged " best regiment at arms " (British regiments) at the 6th Divisional " Assault at Arms," held at Poona from the 23rd to the 28th February, 1908.

The Territorial Force came into being on the 1st April, 1908. The existing Yeomanry and Volunteers were given until the 30th June to transfer to the new force.

In the calendar year 1908, 48,706 men transferred also from the Militia to what was termed the Special Reserve, it being arranged that Special Reservists of the Artillery, Engineers and Infantry were to be drilled on enlistment for six months.

In the infantry the volunteer regiments now became, and were shown in the Army List as, extra battalions of the Territorial line regiments to which they were affiliated, the Militia only having hitherto been shown as extra battalions since the introduction of the Territorial system of nomenclature of line regiments in July, 1881. Another innovation was that the names of all line officers serving at regimental depots were now shown as in the Special Reserve.

In approving of the abolition of the Militia, King Edward VII., on the 21st February, thus thanked them for their services :—

" I take the opportunity of expressing to the Force my " keen appreciation of its services in the past. In peace and " in war the Militia has never been asked in vain to make " sacrifices for the good of the country."

It was also decided that the Territorial force may carry colours on the same lines as in the regular army, whilst a service decoration was approved for the officers and a medal for the men.

On the 9th May, the death occurred of Captain H. S. Logan, of the regiment, from wounds received in action, whilst serving (with the temporary rank of major) with the Egyptian Army.

On account of disturbances which had been caused by the false Mahdi, Abd-el-Kader, in the Mesellamia district, the neighbourhood was visited by the Inspector of the Blue Nile Province, accompanied by a native police officer, both of whom were murdered; whereupon Major Dickinson, Governor of the province, and Major Logan, with a company of the 13th Soudanese Infantry, started for the scene of the murders, a force of mounted infantry and two Maxim Batteries having been despatched from Khartoum. Major Dickinson's force encamped on the night of the 1st May at Katfiyah, to await the arrival of the troops from Khartoum, when Abd-el-Kader attacked his camp, and was repulsed with a loss of 35 killed and 80 wounded, our casualties being :—Killed, 2 Egyptian officers and 8 men ; wounded, Major Logan and one native officer, severely ; Major Dickinson, and Mr. Peacocke (settlement officer), slightly ; and 20 men.

On the 31st July, 1908, new colours were presented to the 1st Battalion at Shorncliffe by Her Grace the Duchess of Rutland, and the ceremony took place in the presence of a large number of military and civilian spectators. His Grace the Duke of Rutland was also present, together with Major-General Utterson, C.B. (Colonel of the regiment) and a large number of the officers' personal friends, all old officers of the regiment and their wives having also been invited. After the ceremony, luncheon was served in the officers' mess, when a presentation, on behalf of the officers, was made by Colonel Burne to the Duchess of Rutland of a diamond brooch in the form of a tiger.

The members of the Sergeants' Mess also celebrated the occasion by giving an " At Home " in the grounds known as Ashley Park, and provided an excellent repast, of which about 1,200 people partook, and the officers gave a ball the same night at the Hotel Metropole, Folkestone, to which about 1,030 guests were invited.

On the 10th September, Colonel G. H. P. Burne retired on retired pay on completion of term of command, when Major L. C. Sherer was promoted Lieut.-Colonel to command the 1st Battalion.

1909

On the return of the 2nd Battalion to Belgaum, on the 14th February from the divisional manœuvres at Poona, the battalion was proclaimed as having won the 6th Divisional shooting trophy.

The laying up of old colours of the 1st Battalion (in use from 1883 to 1908) took place at St. Martin's Church, Leicester, on the 1st July. The ceremony was conducted by the Rev. Canon Sanders, vicar of the parish, and the colours were escorted to the church by an escort of 100 non-commissioned officers and men, under Captain Dent, with Lieutenants Panton and Black of the colour party, accompanied by the band and drums, all of whom had proceeded from Shorncliffe to Leicester by special train on the previous day.

As the unveiling of the County South African War Memorial by Field-Marshal Lord Grenfell took place at Leicester on the same day, the escort of the 1st Battalion (at the conclusion of the service in St. Martin's Church) was marched to the Municipal

Square as a guard of honour, when remarks were heard on all sides on the fine physique and bearing of the men, as well as on the large proportion of medals worn by them.

On the 13th July, the 2nd Battalion was awarded the "Marcks" shield for the best shooting regiment in the Southern Army.

On the 16th December, Brigadier-General the Honourable Montagu-Stuart-Wortley, C.B., &c., made his farewell inspection of the 1st Battalion at Shorncliffe prior to its move to Aldershot, and expressed his deep regret at losing from his brigade "a battalion which had always maintained such an excellent "and enviable reputation for its high standard of efficiency "in all its undertakings, both in work, discipline and games."

1910

On the battalion leaving Shorncliffe, a resolution was unanimously passed by the Cheriton Urban District Council, praising the conduct of the men of the regiment during its three years' stay in the Shorncliffe garrison, which was suitably replied to by Lieut.-Colonel Sherer.

The 1st Battalion left Shorncliffe by rail for Aldershot on the 5th January, in relief of the 2nd Battalion Scottish Rifles, and on arrival was quartered at the Talavera Barracks, South Camp, and joined the 6th Infantry Brigade, 2nd Division.

On the 10th February, the 2nd Battalion was adjudged the "best regiment at arms" (British regiments) at the 6th Divisional "Assault at Arms," Poona.

Major H. L. Croker was promoted on the 11th November to the command of the 2nd Battalion, vice Colonel Bunbury, D.S.O., who retired on half pay, and on exchanging with Lieut.-Colonel Sherer, assumed command of the 1st Battalion on arrival from India.

1911

On the 7th June, the 1st Battalion, whilst training at Frith Hill, was inspected by His Majesty King George V., who was making a few days' stay at the Royal Pavilion, Aldershot.

In consequence of a serious railway strike in England in the month of August, the battalion underwent an unusual experience in being detailed for "strike duty," and, on the 17th of the month, was ordered to London, entraining at the Government Siding, Aldershot, at an hour's notice, each man carrying twenty rounds of ball ammunition. The battalion encamped at Victoria Park (East) until the evening of the 19th, when a move was made to Regent's Park, but, *en route*, orders were received to entrain at King's Cross for York, which was reached at about 4.30 a.m. on the 20th. Five companies were detailed for duty at York Station and on the line, the remainder being in reserve at Fulford Barracks.

On the settlement of the strike, the battalion left York at 9 p.m. on the 25th, and arrived at Aldershot early next morning.

On the 4th February, Major-General Alderson, C.B., Commanding Poona Division, made his farewell inspection of the 2nd Battalion at Belgaum, when he expressed the deep regret of himself and staff and of all units in the Division at the departure from the station of "such a fine regiment," and, commenting on their sporting proclivities, remarked that "a regiment good at games is also good at fighting."

After church parade, on the following day, Major-General Mackenzie-Kennedy, C.B., commanding Bangalore Infantry Brigade, in bidding farewell, endorsed the remarks of high praise by Major-General Alderson, and added: "I have never "met a regiment with whom I would sooner go on service."

The headquarters and right half battalion left Belgaum on the 11th February, and, reaching Madras at mid-day on the 13th, took up quarters at Fort St. George, and the left half battalion proceeded to Bellary.

In the "London Gazette," dated 21st December, it was announced that the King had been pleased to give and grant unto the undermentioned officers employed with the Egyptian Army H.M.'s Royal licence and authority to accept and wear Decorations (as stated against their respective names) which have been conferred upon them by H.H. the Khedive of Egypt, authorised by H.I.M. the Sultan of Turkey, in recognition of valuable services rendered by them:—

. . . Fourth Class, Imperial Ottoman Order of the Osmanieh.—Captain H. D. Beamish, Leicestershire Regiment.

1912

The establishments of the two battalions on the 31st March, 1912, were as follows:—

1st Battalion (Home): 25 officers, 2 warrant officers, 39 sergeants, 16 drummers, 40 corporals, 680 privates; total rank and file, 720; all ranks, 802.

2nd Battalion (India): 28 officers, 2 warrant officers, 45 sergeants, 16 drummers, 40 corporals, 900 privates; total rank and file, 940; all ranks, 1,031.

On concluding this history, on the 31st March, 1912, we find the 1st Battalion under orders to leave Aldershot in the autumn for Fermoy, and the 2nd Battalion under orders to leave Madras and Bellary for Ranikhet.

OFFICER'S SHAKO PLATE (*reduced size*).
1846-55.

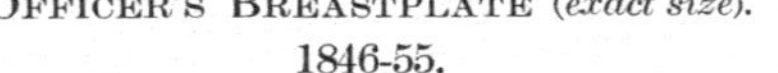

OFFICER'S BREASTPLATE (*exact size*).
1846-55.

CHAPTER XV.

NOTES ON THE UNIFORM, EQUIPMENT, AND THE COLOURS.

1688–1701

AT the earlier period, each company of 100 men in the infantry usually consisted of 30 pikemen, 60 musketeers and 10 men armed with light firelocks. Ten years previously (1678), King Charles II. had added to each of the old British regiments a company of men armed with hand grenades, to be termed the " grenadier company."

Grenadiers appear to have been armed with firelocks, and to have used cartridges, to have had slings, sword-bayonet, and pouch with grenades. They had also hatchets, with which, after firing and throwing their grenades, they were (on the command " Fall on ") to rush upon the enemy's obstacles. Each foot soldier carried a sword and each pikeman a pike, 16 ft. long, each musketeer a musquet, with a collar of bandoliers ; the barrels of the musquets were about 4 ft. long, and carried a ball, 14 of which weighed a pound. All officers of foot carried the spontoon. It had a longer and larger blade than a half-pike, and was rendered more fit for execution by a cross-stop. Officers of the flank companies carried fusils. In 1695, the coats and breeches of the sergeants and soldiers were of a grey colour ; the coats of the drummers purple, with grey breeches. They were also distinguished by badges. Sword-belts were supposed to be of buff leather, but, in reality, as buff was an expensive material, belts (except perhaps in such regiments as the Household Cavalry) were of some other and cheaper leather and of a browner colour. The head-pieces in use during the half-century were of two kinds : the basinet or pott, and the skull cap. The pott was a low-crowned helmet with a brim ; it was sometimes bright and sometimes painted black, and, as a distinction for officers, was ornamented with plumes. Both pikemen and horse soldiers wore, besides the pott, a cuirass, or back and breast-piece, the latter, like the pott, being sometimes black and sometimes bright. By officers of pikemen a sash was worn over the right shoulder.

The following is a copy of one of the earliest orders for the issue of equipment to Colonel Solomon Richards' (17th) Regiment, its date being that on which the regiment sailed from Liverpool to Londonderry :—

" William R.

" Our Will and Pleasure is, that, out of the Stores remain-
" ing within the Office of Our Ordnance, you cause, forthwith,
" to be delivered to Our Trusty and Well beloved Colonel
" Solomon Richards, or such Person or Persons whom he shall
" appoint to receive the same, the several Particulars
" hereafter mentioned :

" Hatchets	30	For the use of the Grenadier.
" Pouches	22	
" Cartouche Boxes	20	
" Primers	63	

" Bayonets, 658, for the use of the Regiment.

" And, for so doing this, together with the usual Indents
" taken for the same, Shall be your Warrant.

" Given at our Court at Whitehall, the 3rd day of April,
" 1689, in the first year of Our Reign.

" By His Majesty's Command,
" Shrewsbury.

" To—The Duke of Schomberg."

1698

On the 28th July, an order was issued for 104 pikes to be delivered to the regiment (8 to each of the 13 companies) in lieu of the like number of " fire lock musquetts " to be returned into store.

1702–1714

All chivalric costume ended in the reign of Queen Anne, when the following was the dress of an infantry officer : Hat, ornamented with feathers, broad brim, two sides of which were turned up, full flowing wig ; square-cut coat and long flapped waistcoat, with large pockets to both ; breeches, tied below the knee, with stockings drawn over up to the middle of the thigh ; shoes ; sword slung over the right shoulder by means of a rich shoulder-belt or baldrick.

Before the advent of the first King George, the crimson silk sash, denoting the rank of commissioned officer, was worn over the left shoulder, and the sword hung in the frog of a leather waist-belt, sometimes placed over and sometimes under the waistcoat. Armour for infantry being now completely thrown aside, the men wore an easy red coat with facings, a cocked hat, breeches with stockings, and a strap below the knee to keep them up.

During the reign of Queen Anne, grenadiers ceased to carry hand-grenades.

1725

In March this year, the King decided that all non-commissioned officers and private men were to wear swords.

1729

By a Royal Warrant, dated November 20th (a re-issue of one issued in 1706), the following were the prescribed "Necessaries for a Foot Soldier," viz. :—A good full bodied coat, well lined, which may serve for the waistcoat the second year, a waistcoat, a pair of good kersey breeches, a pair of good strong stockings, a pair of good strong shoes, two good shirts, two good neckcloths, and a good strong hat well laced.

It is certain that very little information is obtainable as regards the peculiarities which distinguished one regiment from another prior to the year 1742.

1742

In the British Museum is a work entitled "A Representation of the Clothing of His Majesty's Household, &c., in 1742," which depicts a private of the regiment at this period. The coat, very full and roomy, is similar in many respects to that worn in 1714–15, except that the skirts are hooked back, showing the colour of the regimental facing, which was then of the greyish tint shown in Plate 3.

The edgings of the cuffs, lapels, pocket flaps and red waistcoat are trimmed with a distinctive regimental pattern of white lace, with a blue double zigzag on it. As the choice of the lace was left to the colonels of regiments, who provided the clothing, it was not surprising if the pattern changed whenever a new colonel was appointed. A white neckcloth was worn, and the hat was three-cornered, trimmed with white lace. The breeches were red, and white gaiters were worn high above the knee, fastened with dark-coloured garters. The ammunition pouch was supported by a broad leather belt over the left shoulder. There is, unfortunately, no direct evidence of an officer's uniform at this period.

1751

On the 1st July, a Royal Warrant assigned regimental numbers to infantry regiments, and directed the uniform of the 17th to be scarlet, faced and lined with greyish white.

In Windsor Castle is to be seen an oil painting, one of a series, which represents a grenadier of the regiment in a camp scene, showing the cross-belt loosened, which on parade was worn beneath the waist-belt. In other respects the dress varies little from that of 1742, the coat being just as voluminous, but fastening higher, with the addition of a small pouch in front of waist-belt. (See Plate 4.) The imposing mitre cap, with its picturesque embroidery, was made of cloth, the front being of the same colour as the facing of the regiment, with the King's cipher surmounted by a crown. The flap was red, with the white horse and motto over it " Nec Aspera Terrent," a badge that was universally worn by grenadier companies of all infantry regiments. The tuft at top was grey and white, the back part red, with turn-up the same colour as the front, and number of regiment in the centre at back of cap. The lace of the grenadier was white, with the blue double zigzag still retained. In marching order, a knapsack was carried in the form of a bag made of hairy goat skin, and worn over the right shoulder. The uniform of the officers was made up in the same manner as that of the men, laced and lapelled with the colour of the facing, the buttons being set in the same manner as on the men's coats, and waistcoats and breeches of the same colour.

1767

On the 21st September, a warrant was issued for the numbers of regiments to appear on the regimental buttons, which up to this date had been quite plain. That of the officers here shown was made of bone, covered with a fine silver top, handsomely modelled. The private's button was made of pewter.

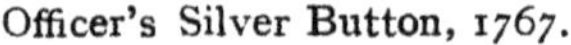
Officer's Silver Button, 1767.

Private's Button of Pewter, 1767.

1768

By a Royal Warrant, dated December 19th this year, the style and cut of uniform underwent a considerable change. The grenadier's cloth mitre cap was replaced by one of bearskin similar in shape, the front of which was a badge of the King's crest, with the motto "Nec Aspera Terrent," the whole in white metal on a black metal ground. Instead of the white neckerchief, a turned-down collar and black tie were now worn, and the coat was scantier, and cut away to show the figure. The waistcoat was shorter, and was changed from red to white. The officers' coats were lapelled to the waist, with the facing of the regiment, and to have cross pockets, and sleeves with round cuffs. The lapels and cuffs to be same breadth as the men's. The patterns of all the regimental laces were altered, that of the 17th being white with two blue stripes and one yellow. The lapels of the officers' coats, three inches wide, were fastened back by silver buttons (white metal for the rank and file) at equal distances, and the loops of lace on the lapels, cuffs and pockets were square-headed and placed at equal distances. Swords having been discontinued by the rank and file, were now worn only by the sergeants and grenadiers. In other respects the accoutrements remained the same. Officers of grenadiers wore an epaulet of silver lace and fringe on both shoulders, and battalion officers on the right shoulder only. Officers of the grenadier company carried fusils, and wore white shoulder-belts and pouches; the other officers carried the esponton, a light steel-headed pike, about seven feet in length, with a small crossbar below the blade. Sergeants carried swords and halberds, the latter, a light ornamental kind of battle-axe, with a long hand or shaft; and wore hats laced with silver, and crimson worsted sashes with a white stripe in the centre. The grenadiers' coats had wings of red cloth on the point of the shoulder, with six loops of regimental lace and a border of the same round the bottom. The brass match-box on the shoulder-belt at this period was probably retained as a distinctive badge only, its use for carrying the fuse to light the grenade having ceased during the reign of Queen Anne. (See Plates 4 and 7.)

Corporals' and privates' coats were laced with the regimental lace, as before described, corporals being distinguished by a silk epaulet on the right shoulder. Drummers and fifers wore red coats similarly lapelled and faced to those of the privates, except that they wore bearskin caps with a plate in front, and a short sword with scimitar blade. Each of the pioneers carried an axe and a sword, and wore an apron and a cap with a leather crown and a black bearskin front, on which was displayed the King's crest, with a saw and an axe in white on a red ground. At the Prince Consort's Library, Aldershot, can be seen an old book, dated 1768, showing the privates' uniforms of every British infantry regiment at this period.

1769

From the Inspection Report of the regiment, dated Chatham, 17th May, it would appear that the greyish-white facing, authorised in 1751, had now got to be officially recognised as white. The following extract from it states: "Officers' "uniforms, shown as faced with white, white waistcoats and "breeches, silver epaulets, laced hats. The grenadiers have "cloth caps, the furred ones not being yet finished. Gaiters, "black linen with leather tops, and black garters neatly "made."

By the new Clothing Regulations, sergeants of infantry regiments were to carry fusils and pouches, and the annual allowance of powder for each regiment was increased from eight to ten barrels.

1771

Soldiers of the light company, which was added this year, wore short jackets, red waistcoats, and a leather cap, almost as small as a skull cap, with a large round peak, straight up in front; their waistbelt had two frogs, one for the bayonet, the other for the hatchet, when the soldier was on duty; at other times on the march it was to be tied upon the knapsack, and the light company gaiter was to be " as high as the calf " of the leg and no higher."

1775

In the Inspection Report of the regiment, dated 9th June, the officers' uniforms are described as: " Scarlet, small round " cuff, collar and lapels to the waist white, with a silver laced " epaulet, silver buttons numbered, white lining, white waist-" coat and breeches, silver hilted swords, crimson and gold " sword-knots, silver gorgets, everything according to His " Majesty's order."

1776

By the year 1776, it became the universal custom to wear the white leather shoulder-belt for the sword outside, instead of inside the coat, when regiments, disliking the plain buckle and tip, which had hitherto been conspicuous on it, began to adopt something more ornamental in the shape of a small silver plate. For some years the design on this oval plate was engraved with merely the Royal cipher " G.R.," and the crown and regimental number on it (see Plate 10), representing probably the earliest breastplate worn by the officers of the 17th Regiment. An original of this was found about the year 1898 whilst digging the foundations for some houses in the city of New York, where the regiment was frequently stationed at intervals during the American Revolutionary War.

By the year 1799, the belts had become wider, and in many cases the ornaments, instead of being engraved, were raised upon the plate, much enhancing the effect. (See Plate 10.)

1784

An Adjutant-General's order, dated 20th March, directed that belts of infantry soldiers were to be worn over the right shoulder as cross-belts to support the bayonet, instead of round the waist as hitherto.

1786

In accordance with a General Order, dated 3rd April, infantry officers were to lay aside the esponton, and to provide themselves with a strong, substantial uniform sword, with straight cut-and-thrust blade, an inch broad at the shoulder and thirty-two inches in length. The hilt, if not of steel, was to be either gilt or silver, according to the buttons on the uniforms. The sword-knot was to be of crimson and gold, in stripes.

1789

By General Order, dated 27th March, halberds were abolished, and swords substituted for the sergeants. Brass drums also took the place of wooden ones.

1791

On the 6th October, it was directed that all field officers of cavalry and infantry were to be distinguished by wearing two epaulets. Grenadier and light company officers, who had worn two for some time, were now to wear a grenade and bugle respectively on the epaulet.

On the 6th October, halberds were replaced for sergeants by pikes, two being sent to the regiment for trial and report, which were approved for use in April the following year.

1795

An order, dated 19th July, directed that the use of hair powder was to be discontinued by non-commissioned officers and men, but not by officers. Officers' silver gorgets were this year replaced by those of gilt metal, which were of universal pattern and worn up to 1831.

By a Royal Warrant, dated 1st September, the following was the schedule of necessaries for each soldier of infantry, viz.: three shirts, two pairs shoes, two pairs stockings or socks, one pair of long gaiters, one forage cap, one pack, one stock, one black ball and two brushes.

1796

By the Warrant of the 1st May, the coats for all ranks were fastened down to the waist, by which the sleeved white waistcoat (which afterwards became an undress garment for the rank and file) was completely hidden. The lapels were continued down to the waist in such a way as to make a double-breasted coat, with a high stand-up collar to admit of the large neckcloth then worn. The jacket for the rank and file was single-breasted, with ten buttons, and loops of regimental lace across the chest.

1799

An order was issued this year for officers and men of infantry regiments (except those of flank companies) to wear their hair queued ten inches long, including one inch of hair to appear below the binding and to be tied a little below the upper part of the coat collar.

1800

By a General Order, dated 4th February, the cocked hat worn by the men was discontinued, and a cylindrical shako with a peak was introduced. It was of lacquered felt and ornamented with a large brass oblong plate of universal pattern, bearing the King's crest, with a red and white tuft in front rising from a small black cockade, and was worn without a chin-strap.

1801

A Warrant was issued in April directing that all men of the Foot Guards and infantry should be provided with a serviceable greatcoat.

1802

A Horse Guards letter dated 1st July, directed epaulets and shoulder-knots to be discontinued by non-commissioned officers of the Foot-Guards and Infantry and, in lieu, chevrons made of lace used in their regiments: sergeant-majors four; sergeants three, and corporals two upon the right arm.

1806

In October, a black felt shako was substituted for that of lacquered felt, the brass plate of which was smaller in front and more oval than that of 1800; it was surmounted by a crown, and the King's cipher, with number of regiment immediately below it. It had a red and white worsted tuft and cockade at the side, and suspended across the front was a crimson and gold twisted cord with tassels, the same cord in white for the men of battalion companies, and green for those of the light company. This shako was worn without a chin-strap by the infantry throughout the Peninsular and Waterloo campaigns. (See "Officer, 1814," Plate 11.)

1808

A General Order, dated 20th July, directed that the use of queues be dispensed with in the army, care being taken that the men's hair was cut close to their necks in the neatest and most uniform manner.

1811

Infantry officers were authorised to discontinue wearing cocked hats and to wear a head-dress similar to that of the men, a double-breasted red jacket with very short skirts, and a grey overcoat and grey trousers, the same as the men.

1815

A General Order, dated 2nd August, authorised the introduction of a bell-shaped shako, eleven inches in diameter on the top, and seven and a half inches deep, with silver chain scales, which were allowed to be fastened up in front, below the black cockade; upright white feather twelve inches long, shako-plate silver (see Plate 13), star oblong 3½ by 3⅛ inches; silver lace (2½ inches), oak-leaf pattern round top, and regimental half-inch vellum lace at bottom. Grenadiers retained their bearskins with silver tassels for the officers and white for the men. Green feather and bugle badge for light company.

It would have been about this year, that the officers of the regiment adopted for the first time an oblong shaped breastplate (Plate 12), to replace the silver oval one taken into use in 1799 (Plate 10). The new one measured $3\frac{3}{4}$ by 3 inches, and consisted of a dead gilt plate, wreath and crown, the two last picked out with bright gilt; and had rounded corners. A star inside wreath, the long rays of dead silver, and the short rays of burnished silver. In centre, gilt garter, and numeral "17," the latter on a green enamel ground.

The Black Stripe in the Silver Lace.

Lacemakers' books show that in 1822 the officers' regimental silver lace had a black stripe through the centre of it, the origin of which (as mentioned in the "Preface") is obscure, though tradition has handed down that it was worn as an emblem of mourning for General Wolfe. The late Mr. S. M. Milne, of Calverley House, near Leeds (who up to the date of his death, 16th May, 1910, was considered the greatest authority of the day on such matters), wrote: "I can well "believe that the 17th Foot would be quite the oldest and "first corps to wear this black stripe in the officers' lace. A "black stripe was introduced into the officers' vellum lace "towards 1790, simply on account of brightening the effect, "and for no exact reason such as mourning. It was adopted "by many line regiments *before they ever fired a shot*, and by "quite a number of militia regiments also."

1820

This handsome skirt ornament consisted of a richly spangled silver star, with gold embroidered "17," on a raised silver lace medallion.

Attention was called this year, to the fact that the gorget formed a part of the officers' equipment, and must be worn when on duty.

Short coats were abolished for all ranks, and, in 1822, breeches and leggings.

Officer's Silver Skirt Ornament, 1820.
(Reduced to half size.)

1823

In June, bluish-grey trousers replaced the darker grey ones and were worn until 1830. (*Vide* Sergeant, Grenadier Company, Plate 16.) White linen trousers were also introduced for summer wear (between the 1st May and 14th October). (*Vide* Light Company Bugler in same Plate.)

1825

The badge of the "Royal Tiger," superscribed "Hindoostan," having been conferred by King George IV. on the 25th June this year, and oblong breastplates having replaced those of oval shape, the officers of the regiment decided to have a new silver breastplate and shako plate (Plates 13 and 17), together with the waistbelt clasp, (shown at Plate 30). The breastplate was of solid silver, 4 by 3¼ inches; on it, a silver diamond-cut star; the tiger and "Hindoostan" dead gilt; the number in dead matted gilt. The shako star, silver, diamond cut; the tiger and scroll, dead gilt, and the numeral, dead matted gilt.

In an old War Office book, a letter from the Adjutant-General, dated April 12th, 1825, shows that the regiment had just applied for a "change in the pattern of the facing," and that the Commander-in-Chief did not approve of the alteration. From this it may be inferred that there was no desire to change its colour from white, and that the proposed alteration referred, most probably, to the pattern of the laced loop then worn on the coat collars. It is regretted that the letter, submitting the application from the regiment, is not forthcoming in the War Office books.

1826

The private soldier's coat was altered in cut, the lace removed from the collar and a single loop placed on each side, the loop across the chest made broad at the top, tapering narrower towards the bottom, and the lace removed from the coat skirts, except the loops on the slashed pockets. Officers' rank was distinguished by the epaulet, according to the instructions in the General Order of February, 1810. Field officers wore two epaulets, a colonel having a silver star and crown embroidered on the strap; lieut.-colonel, a silver crown; major, a silver star; whilst captains and subalterns wore a single epaulet on the right shoulder.

1828

The Inspection Report this year, dated 13th November, shows that the Inspecting General found fault with certain articles of the officers' uniforms as being unnecessarily costly in price, thus deviating from the Horse Guards' Circular on the subject, which had been issued in the previous month. Exception was taken to the following, viz.: 1. The handsome silver breastplate and shako star, which had been adopted by the officers from the 13th October, 1825. 2. Silver laced trousers for dress. 3. A forage cap for the barrack yard (not yet authorised). 4. A silver dress waistbelt, entirely at the option of officers.

In December, another change was made in the shako. Whilst it remained bell-shaped, the lace was to be stripped from it, its height reduced to six inches, and a large gilt star plate worn in front. (See Plate 14.) It had gilt scales to fasten under the chin, and with it was ordered a rich gold festoon, with cap lines and tassels of gold. (See centre figure in Plate 18.) The same was made of worsted for the men, white for battalion companies and green for light companies, and with it was worn a plume of white feathers twelve inches high.

1829

Exception having been taken to the silver breastplate, the same pattern was now adopted in gilt. With a view to preventing an excessive amount of lace being worn on the officers' coats, the Warrant of February this year introduced the well-known coatee of universal pattern, which with little alteration was worn by officers until after the Crimean War. Of the two silver buttons here shown, the larger one is of a much earlier date. In the smaller type we have represented, for the first time, a regimental button showing

THE PERPETUAL OR UNBROKEN LAUREL WREATH,

to which it has been suggested that some glory was attached.

Officers' Silver Buttons before 1830.

The regimental "Digest of Service," however, of the 1st Battalion (dated "Hull Citadel, 1st May, 1824"), makes no allusion to any peculiarity, either in this button or to the black line in the officers' lace previously mentioned. No regulation button design existed in the Army until the introduction of the Territorial system in July, 1881; it was therefore optional with commanding officers to choose their own designs. There is no record of a perpetual laurel wreath having been specially conferred on any of the appointments at this period, with a view to commemorate any special engagement that had taken place, but there seems reason to believe that the perpetual laurel wreath was specially chosen on the button to perpetuate the memory of the regiment having covered itself with glory in 1777, when, under Lieut.-Colonel Mawhood, it charged through the right wing of the American Army.

By a circular letter, dated 21st March, 1829, a forage cap for officers was authorised for the first time, of blue cloth, with a large flat stiffened top and peak, having only a band of cloth round it of the colour of the regimental facing, and no number.

The coatee now introduced was scarlet, with collar and cuffs of the regimental facing, Prussian collar with two loops and uniform buttons at each end. It was double-breasted and the buttons (for the 17th) placed in pairs, with intervals between them. White cuffs, with a scarlet cuff slash, and on it four buttons and laced loops, cuffs 2¾ inches, slashed flap on skirts with four loops and large buttons, white kerseymere turnbacks and skirt lining, regimental skirt ornaments, those of the rank and file consisting of a regimental button for the battalion companies, and of a metal grenade and bugle-horn respectively for the flank companies. All ranks of officers now wore two epaulets, a universal pattern having been adopted for the whole of the infantry. An example of the single epaulet worn prior to this date is shown at Plate 9.

1830

According to Regimental Standing Orders dated Chatham, 14th January, gorgets were worn on all duties that were performed in red. The coatee to be worn at all parades, regimental as well as general, whether with or without arms, and on all garrison duties with the white shoulder-belt and sash. It was always to be worn at mess and at evening parties with a black leather belt beneath the coat, and a sword, but no sash.

Officer's Gilt Button, 1831.

Silver lace and buttons were now replaced by gold lace and gilt buttons for officers. The officers' gilt button now introduced continued to show the perpetual or unbroken laurel wreath, but this pattern for the regiment was not exclusive.

The new pattern buttons at this period of the 26th (Cameronians) and 28th Regiments also showed unbroken wreaths, which, however, differed in design from each other and from that of the 17th. The Warrant of August this year made several changes. Trousers of "Oxford mixture" were substituted for those of bluish grey; shako lines and tassels were abolished, plumes shortened to eight inches, and a green ball-tuft introduced for light infantry and light companies. (See figure on right, Plate 18.) Sergeants of infantry were armed with fusils instead of pikes. A red fatigue jacket succeeded the white one hitherto worn by the rank and file, the colour of the collar and cuffs (as well as the chevrons for sergeants) to correspond with the facing of the regiment, and finally the gorget was abolished.

Up to this period the dress of bandsmen does not appear to have been taken seriously by the authorities, bandsmen having hitherto been dressed in varied and fantastic costumes. (See Plate 15.) The drum-major and bandsmen here represented (from a drawing by E. Hull, a well-known artist of that period) are shown as wearing pink trousers, and, in some regiments, bandsmen wore scarlet. It was however authorised this year that bandsmen were to wear white coatees as here shown. In this illustration the instrument held by the bandsman was known as the "serpent," a few old specimens of which can still be seen in London, and there is also one in possession of the Sergeants' Mess of the Depot at Leicester. This instrument was in use from 1820 to 1830, when it was replaced by one called the "ophicleide," which had a similar term of ten years' existence, and was succeeded by the present euphonium in 1840.

1831

A new breastplate was this year taken into wear by the officers. (See Plate 17.) It deviates from the original order, conferring on the regiment the badge of the "Royal Tiger" in having the word "Hindoostan" beneath the tiger instead of superscribed, and it was slightly smaller than the breastplate which preceded it. It consisted of a silver diamond cut star on a plate of dead gilt, silver crown and matted silver numeral. The tiger dead gilt; "Hindoostan" bright gilt letters.

On the 18th May, field officers of infantry were ordered to discontinue wearing the shoulder-belt with slings, adopting instead a white leather waistbelt with a gilt clasp in front. (See Plate 30.) Instead of the leather scabbard they were also directed to wear one of brass; adjutants, at the same time, being directed to wear steel scabbards. Spurs for mounted officers were to be of yellow metal with necks 2½ inches long, including rowels.

1833

In January, the narrow welt of red cloth down the outer seams of infantry trousers, as at present worn, was authorised for the first time.

1834

A new officer's forage cap was introduced of blue cloth, with a black silk oak-leaf band and the regimental number, one and a half inches deep, embroidered in gold on the front. The cap had a peak, and according to Regimental Orders the men's brass numbers had to be placed one inch above it. The undress single-breasted blue frock coat now presented a handsome appearance, being ornamented with scales, which consisted of blue cloth shoulder-straps, laced with broad regimental gold lace, and terminating with gilt metal crescents.

The white plumes in the shakos were replaced by white worsted ball-tufts (see Plate 22), the light company wearing green tufts.

1836

By General Order this year, the coloured regimental lace, so long worn by the rank and file, was abolished, and replaced by plain white tape lace, but the mode of wearing the square-headed loops across the chest at equal distances was retained. The sergeants were directed to wear double-breasted coats without any lace on the chest, with white epaulets and wings for those of flank companies. The coloured lace, however, of the drummers of regimental pattern which had been worn since about the year 1820 was continued.

1839

The brass shako plate, with the regimental number, of universal pattern worn by the rank and file, was this year replaced by a round brass plate three inches in diameter, with a crown above it and the number raised in the centre.

1840

The new percussion muskets were generally introduced into the army.

In accordance with the fashion at this period, gilt buttons were largely worn in plain clothes, noblemen and gentry wearing them of various patterns, whilst those holding official appointments wore them of a special design. Military officers accordingly adopted buttons of a special regimental pattern, of which this was the design chosen by the officers of the 17th Regiment, and it was from this that a smaller button came later into use for the officer's mess waistcoat. (See page 255.)

Officer's Gilt Button, 1840.

1841

On the 7th October this year, an application from the officers was made, through the Colonel of the regiment (General Sir F. A. Wetherall, G.C.H.), for permission to wear on their shakos and breastplates the motto :—

" VENI ET VICI."

The cause for its adoption cannot be explained, as the letter making the application is unfortunately not obtainable in the old War Office records, the reply only being forthcoming, dated Horse Guards, 7th October, 1841. It expresses the regret of the Commander-in-Chief (Lord Hill) at not being able to forward the recommendation to Her Majesty, on the ground that there was " no precedent for permitting a regiment " to assume a motto, such as that proposed for the 17th Foot." It appears, however, that in anticipation of the motto being granted, orders had been issued for the manufacture of a new breastplate and shako plate, bearing the motto (see Plate 23), which on receipt, in September, 1843, were in use up to October, 1845. The breastplate was of burnished gilt, four by three inches, scroll dull gilt, and garter, tiger, and number, silver. The shako plate all of the same coloured metal; the centre and points of star bright burnished gilt. The crown and ornaments all of dead gilt, and letters on scrolls burnished.

1845

Bearskins were abolished for the grenadier companies of infantry regiments, and a new shako was introduced known as the "Albert" pattern, six and three-quarter inches high, with a patent leather peak in front and a narrower one behind; officers had a gilt star plate, surmounted by a crown, bearing within a wreath formed of a laurel and palm branch, a garter inscribed with the title of the regiment; in the centre the number, the whole of the same metal, without any difference in colour. Also a gilt chain, fastening at sides, with rose pattern ornament. A ball-tuft completed it, two-thirds white at top and one-third red, for battalion and field officers, all white for grenadiers, green for light companies. The men retained the brass plate they had worn with the large-topped shako. (See Plates 22 and 24.)

In October this year, a new shoulder-belt plate was taken into wear of gilt matted or frosted metal, nearly covered with a silver star, having a silver crown above it; one silver star ray to the right bearing " Khelat," one to the left with " Ghuznee," and one from the centre below with " Afghanistan "; just under the crown a silver horizontal scroll with

"Hindoostan"; in the centre of the star a gilt garter, having upon it "Leicestershire Regiment"; within the garter, upon a raised silver centre, a silver tiger above the numerals 17, also in silver. It was the last plate worn, on the abolition of the coatee, in the "Dress Regulations" of the 1st April, 1855.

A new waistbelt clasp was also introduced this year. (See Plate 30.)

The red and white striped sashes worn round the waist by the sergeants of the regiment were discontinued, and a plain crimson one two and a half inches wide substituted.

1848

By a General Order dated 30th June, the laced loops and buttons on the skirts of the officers' coatee were abolished, leaving only the skirt ornaments. The blue frock coat with shoulder scales was also discontinued, and a plain shell jacket, with collar and cuffs of the regimental facing, was introduced. A black patent leather sling sword-belt with snake clasp was worn with it, and a grey greatcoat was also taken into use by officers in lieu of the blue cloak.

1850

A plain shoulder belt, without breastplate, to carry the pouch, was authorised to be worn in the ranks, the bayonet being hung on a frog from a waist belt.

1855

On the 1st April, the coatee had given place to a long double-breasted tunic, with lapels (for officers), which were made to fold down at the top and show the white lining, but the lapels had to be buttoned over on parade or duty. This pattern tunic did not find favour long, and was succeeded, two years later, by the single-breasted tunic of to-day. Slashes of lace, with buttons, were worn on the cuffs, and similarly on the skirts. The crimson sash was worn over the left shoulder, retained by a twisted cord of crimson silk, and the same, with the blue frock coat, in undress. There were then no collar ornaments, beyond an officer's rank, which was designated on the collars of the tunics only, except field officers, who also wore it on the collars of their frock coats.

The respective ranks were shown as follows :—

Ensign, a silver embroidered star.
Lieutenant, a silver embroidered crown.
Captain, a silver embroidered crown and star.
Major, a gold embroidered star.
Lieut.-Colonel, a gold embroidered crown.
Colonel, a gold embroidered crown and star.

Trousers, with the scarlet welt, were of the same pattern as the present day, dark-blue serge being worn from 1st May to 14th October, and Oxford mixture cloth from 15th October to 30th April.

The shako was of the "Albert" pattern, with black leather chin-strap and peaks front and back, and lighter in make than its predecessor, the first double-peaked shako, introduced in 1846. Two rows of regimental lace were worn round the top of it by lieut.-colonels, and one row by majors only.

The shako plate, of this date, was an eight-pointed star (3⅝ inches in extreme diameter), surmounted by a crown, having the regimental number in bright gilt on a black leather ground, inside a garter proper. (See Plate 26.)

A white enamelled leather belt with new clasp (see Plate 30) was worn outside the tunic and the frock coat, and sword scabbards were of black leather, with gilt mountings, except brass for field officers and steel for adjutants, and at a later period, when regimental instructors of musketry were appointed, they also wore steel scabbards. The sword knot was of gold embroidery, with crimson stripes.

The band wore white tunics, double-breasted, with scarlet facings and a red piping up the back and sleeves. A double-breasted blue frock, for undress, had been authorised for officers, and a new blue cloth forage cap with a straight peak showing the regimental number embroidered in front. The white tape lace in the ranks now disappeared, lace of regimental pattern being retained by the drummers, whilst white piping was introduced for the first time, and brass buttons, for the rank and file, replaced the old pewter ones.

The button now introduced was larger and of this pattern.

Officer's Gilt Button, 1855.

1857

With the new issue of clothing this year, single-breasted tunics were introduced.

1858

On the abolition of the flank companies in 1858, the white and green shako ball-tufts, specially worn by them, disappeared, the whole regiment now wearing these alike, viz.: upper part $\frac{2}{3}$ white, and lower $\frac{1}{3}$ red.

1859

The coatee, succeeded by the tunic, had, up to now, been the authorised mess dress for infantry officers, but for many years it had been customary to wear the shell jacket open with a waistcoat, the pattern and material of which were decided solely by officers commanding regiments. The wearing of the shell jacket at mess was authorised by Circular Memorandum, dated 9th June, this year.

Of the two buttons here shown, that of the earlier date was the first worn with the mess waistcoat, its pattern being taken from the button shown under the year 1840.

1858. 1903.
Mess Waistcoat Buttons.

1861

This year saw the introduction of a new shako of blue ribbed cloth, single-peaked, and smaller and lighter. The plate was a gilt star of eight points surmounted by a crown, and having the number of the regiment cut out, within a garter. This shako was worn with a black leather chin-strap. (See Plate 26.)

1866

Officers' black leather scabbards were replaced by steel ones, field officers retaining theirs of brass.

1867

An officer's blue patrol jacket replaced the blue frock coat, and pioneers' white leather aprons were discontinued.

1868

The slashed cuff on the tunic was abolished, and pointed cuffs, with distinctions in the amount of lace for various ranks, were introduced. A levee dress was instituted, consisting of gold laced trousers, and a crimson and gold sash, and sword-belt, all of universal pattern. Authority was received this year for the regiment to wear the badge of the "Royal Tiger" above the number on the forage cap.

1871

Early this year, a new shako, ornamented with narrow gold braid and a gilt chain, was introduced for officers, that for the ranks being without the braid. The old star plate was replaced by a garter with number inside, surrounded by a laurel wreath, the whole surmounted by a crown, and it was the last pattern of shako worn by British infantry. (See Plate 29.) The men's tunics were changed, in colour, from red-brick to scarlet, and a Glengarry cap introduced as a forage cap. The 2nd Battalion, then quartered at Aldershot, also received the valise and Wallace equipment, the old knapsack being withdrawn. In September, white cloth clothing was abolished for bandsmen, who were now designated by a worsted badge of crossed trumpets, on the right arm of their scarlet tunics, this badge being later represented in metal. Scarlet serge frocks also replaced the red-brick coloured cloth shell jackets for the men, and, at the same time distinctive regimental patterns of drummer's lace were finally abolished, having been gradually discontinued since 1866. Thus there disappeared the accompanying fringe, also of red, white, black and yellow (the same colours as the lace), which the drummers had worn for about fifty years. These were replaced by a smaller red and white fringe, and white lace with red crowns, both of universal pattern. November, this year, saw the introduction of a button in the ranks, of universal pattern, showing the Royal Arms, and without any regimental numbers.

1872

Owing to the great variety of mess uniforms that now prevailed throughout the infantry, one of universal pattern was introduced, fastening at the neck, and both jacket and waistcoat edged with teat buttons.

1878

A cork helmet, covered with blue cloth, with a gilt spike and chain, replaced the shako as a headdress for the infantry, the plate being gilt, with number of regiment in the centre within a garter bearing the Royal motto, a laurel wreath round the garter and a crown above, and was taken into wear in the following year. (See Plate 29.)

1880

Officers' badges of rank were removed from the collar and placed on the shoulder cords of the tunic and mess jacket; rank was also similarly shown on the cloth straps which had been added to the patrol jacket.

1881

With the advent of the Territorial system, from the 1st July, and the abolition of regimental numbers, there was a general upheaval of everything referring to regimental distinctions, which, however, was less marked in the first twenty-five double battalions of the line than in the single battalions which followed, on account of the indiscriminate linking which the latter underwent.

The regiment (now "The Leicestershire") was fortunate in retaining its white facings, which, under the new system, had become the established colour for all English (except Royal) regiments.

A new officers' forage cap was now introduced, with drooping gold embroidered peak, the first Territorial badge consisting of a gilt star with silver ornaments—the Royal Tiger, superscribed "Hindoostan," above the Harp of Ireland, the latter having been conferred on the Leicester Militia for meritorious service during the Irish Rebellion in 1798.

The black stripe in the centre of the officers' gold vellum lace now disappeared, and the lace itself was replaced by gold lace of "rose" pattern, universally worn by all English regiments, the former black stripe through the centre being now represented by two narrow black threads at the sides. Shortly after, however, the black stripe was inserted in the officers' gold shoulder cords, a distinction which has since been extended to a few other regiments.

1882

Saddle cloths, hitherto used in review order by infantry mounted officers, were discontinued.

1887

Brown gloves first came into wear, and, except in review order, became universally worn from the year 1900.

1893

In May, a field service cap of a special pattern came into use.

1900

The sabretache for mounted officers, steel chain reins, and bearskin cover to wallets were abolished, also the levee dress.

1902

Throughout the South African War, infantry greatcoats were mostly of khaki coloured cloth, and, at its conclusion, numerous changes in dress took place. Brass scabbards and brass spurs were replaced, for field ranks, by those of steel, and the greatcoat was to be carried on the saddle. The following articles were introduced: a double-breasted blue frock coat for officers; a new forage cap for all ranks, with a large flat stiffened top and sloping peak; a new pattern skirt to officers' tunics, with a new sword belt, now worn beneath the tunic and the sash worn round the waist.

From this date, the khaki service dress was universally worn as undress uniform on all parades and regimental duties, with putties and the Sam Browne belt.

1909

The web equipment of 1908 pattern was taken into wear by the 1st Battalion from the 7th October this year.

1911

The officers, 1st Battalion, discarded, this year, the white pipe-clayed colour-belts which had been hitherto supplied from Pimlico, and replaced them regimentally with two very handsome broad belts of gold lace, with black stripes, bearing a suitable gilt breastplate, displaying the Royal Tiger.

NOTES ON THE COLOURS.

Colours and standards have been in use by regiments of the British Army, ever since its formation, as a permanent force, which may be said to date from 1661, in the reign of King Charles II.

Their patterns, and the devices on them, appear to have been optional, for many years, with Colonels of regiments. Important instructions, however, about clothing and colours were delivered to the Clothing Board by Robert Napier, Adjutant-General, on the 11th November, 1749.

" No Colonel was to put his arms, crest, device, or livery, " on any part of the appointments of his regiment."

" No part of the clothing or ornaments of the regiment " was to be altered after the following regulations were put " in execution, except by His Majesty's, or H.R.H. The Duke's " (Cumberland) permission."

The regulations affecting colours were as follows :—

The first colour of every marching regiment of foot is to be the Great Union. The second colour to be the colour of the facing of the regiment, with the Union in the upper canton, except those regiments which are faced with white or red, whose second colour is to be the red cross of St. George in a white field, with the Union in the upper canton.

In the centre of each colour is to be painted in gold, Roman figures, the number of the rank of the regiment, except those regiments which have royal devices or ancient badges; the numbers of their rank are to be painted towards the upper canton. The length of the pike, and the colour itself, to be the same size as those of the royal regiments of Foot Guards, six feet six inches flying, and six feet deep on the pike ; length of pole, spear and ferril included, being nine feet ten inches. The cords and tassels of all colours, to be crimson and gold. The camp colours to be of the facing of the regiment, with the rank of the regiment thereon.

The following dates have been traced as those on which sets of colours have been issued to the 17th Regiment :—

OFFICERS IN THE CRIMEA, 1855.

Standing Lieut. Boyd, Capt. M'Kinstry, Capt. Gordon, Lieut. Traherne, Capt. Colthurst, Lieut. Lukin, Brigdr.-Genl. M'Pherson, Lieut. Tompson, Lieut. Lees, Lieut. Williams, Capt. J. Croker

Sitting Capt. Heigham, Lieut. Smyth, Capt. O'Conor, Lieut. Cecil M'Pherson

From the Inspection Returns (at the Record Office, Chancery Lane, W.C.), it appears that a set was received in 1766.

These colours were lost at Stoney Point (a small fort on the Hudson River, above New York), on the 16th July, 1779, when this post was besieged by an overwhelming number of Americans.

The regiment arrived home after the war, in August, 1786, and was actually without any colours, until the receipt of a new set, at Chatham, on the 14th April, 1787; an unusual and unheard-of thing.

1799

The formation of a second battalion to the regiment in August this year, took place so hurriedly, for it to take part in the Helder expedition, that not only was there not time to clothe many of the volunteers to it in the uniform of the regiment (in consequence of which many arrived in Holland in their militia clothing), but also, for the same reason apparently, no colours were issued to the battalion; nor is there any record of its having received any colours during its term of existence up to August, 1802.

1802

In 1802, in consequence of the Union with Ireland, the colours of the whole army were altered, *i.e.*, the shamrock was introduced into the Union wreath, and the red saltire cross of St. Patrick placed upon the white saltire cross of St. Andrew.

The Heralds' College books state that a new set was given out in November, 1802, when the regiment was at Limerick, and the following extract from "Billinge's Liverpool Advertiser," dated December 13th, 1802, shows the fate that befel the previous set of colours which had been issued at Chatham in 1787.

That journal states: "On the night of the 17th ultimo, "a vessel was wrecked at the Red Strand, about three miles "from the town of Clonakilty, in Ireland, and every person "perished. It appears, from part of her stern which came "ashore, that she was the 'Mary' of this port. . . . Some "of the bodies of the crew were washed ashore near Rose

"Carbery, as were also the old colours of the 17th Regiment "of Foot; these are conjectured to be the old colours of the "regiment, which were being sent to England from Limerick, "where the 17th are now quartered."*

1830

The next set was issued in 1830, at Chatham, prior to embarking for New South Wales, and, on both colours now received, there was displayed for the first time,

"THE GREEN TIGER,"

as can be seen (December, 1910) by the heraldic devices on the old flags in St. Martin's Church, Leicester; the form of tiger on each colour (with tail above the back), being similar to that shown in Richard Cannon's plate, dated 1848,† represented at page 14, there having been no further issue of colours to the regiment until early in 1854. The origin of the tiger being represented as green with gold stripes is shrouded in mystery, seeing that, in the few other regiments on whom the emblem was conferred before 1825, the badge had always been displayed as a tiger "proper." Neither the War Office, the India Office, nor the Heralds' College have been able to throw any light on its origin, the only information obtainable from the latter, being that the Green Tiger, as displayed by Richard Cannon, is the exact type of the emblem adopted by the Sultan Tippoo, who fought against the British at Seringapatam; and in the Heralds' College books it is heraldically described as: "A Royal Tiger, passant-gardant, vert, striped and spotted or." This same Seringapatam trophy has also been introduced into the arms of the Marquis of Wellesley and other Indian celebrities. Tradition has handed down, regimentally, the most probable explanation of its origin on the regimental colour, viz., that in one of the actions in India in which the regiment was engaged, between 1804 and 1823, a standard was captured from the enemy, on which was borne a green tiger, and this fact having been communicated to the Inspector of Colours,

* Contributed by Mr. W. Thomas, late Leicestershire Regiment.

† It is worthy of note, that the width and positions of the white stripes, in the King's Colour, as shown by R. Cannon (herein reproduced as Plate 2), are technically incorrect.

the tiger was reproduced in that form, either on the representation of the officers or owing to a whim on the part of the Inspector of Colours then in office.

The wording of the grant describes the Royal Tiger having been conferred, on the 25th June, 1825, "as a lasting "testimony of the exemplary conduct of the corps, during "the period of its service from 1804 to 1823." During that interval, the regiment had served in Bundelcund in 1807, on the Sutlej, 1808, on the Nepal frontier, 1814–15, and the operations for the relief of Nagpore in 1817.

It has been represented that, in the action with the Nepalese (1815), a standard was taken by the regiment, which bore an emblem of a green tiger, but of this there is no direct proof.

Another conjecture is, that perhaps a similar emblem was on a standard which may have been captured by the 17th, when the regiment drove off the Arab infantry with great loss at Jubbulpore in 1817; in that case, it would be a Mahomedan emblem, and, in all probability, similar in design to that on Sultan Tippoo's standard, but as no official record is obtainable regarding it, nothing definite is known.

Beyond the badge of the "Green Tiger" having been approved for the regimental colour, by the Inspector of Colours, in the early part of the 19th Century, it seems evident that as a regimental badge in this form, no importance has been attached to it by the authorities, no directions having ever been given that it should be produced as "green with gold "stripes," either in embroidery, enamel, or metal of any kind, on any of the regimental appointments.

The following is a description of the colours received in 1830:—

The Royal, or First, Colour bore, in the centre, the Imperial Crown. Below it a wreath, in two parts (crossed at top and bottom), and, within this, the Green Tiger, above the number "XVII.," the latter enclosed in two circles, containing the word "Leicestershire." On the left of the colour (next the staff), reading from the top, were inscribed: "Afghanistan," "Ghuznee," "Khelat," "Hindoostan."

The Second, or Regimental Colour, bore the red Cross of St. George, in a white field, with the Union in the dexter canton.

On the left arm of the cross (next the staff) was inscribed "Afghanistan," on the right arm "Ghuznee," on the lower arm "Khelat," on the upper arm the Crown, above the Green Tiger, superscribed "Hindoostan," which, in turn, is above "XVII.," enclosed in two circles, containing the word "Leicestershire," surrounded by the Union wreath. In both flags the numeral and county title are in blue letters on a red ground. The poles were surmounted by a metal design in the form of a pike-head, as shown in Plate 2.

1840

In an old War Office book is a copy of an Adjutant-General's letter, dated Horse Guards, 26th October, to the Colonel of the regiment (General Sir F. Wetherall, G.C.H.), in reply to an application that had been made on its behalf, to assume the words 'Prince Town' on the colours. The letter stated that: "the rule laid down, in according badges "and distinctive marks to the colours and appointments of "regiments for service in the field, does not recognise claims "of this nature, arising out of events that had occurred as far "back as the year 1777, and that Lord Hill cannot, therefore, "feel justified in recommending Her Majesty to permit the "17th Regiment to assume the words 'Prince Town' upon its "colours, although the gallant conduct of that regiment in "America, on the occasion specified in Lieut.-Colonel Croker's "letter to you, is recorded by Sir William Howe and Lord "Cornwallis in their despatches."

It is regretted that the original letter, showing the application from the regiment, is not forthcoming amongst the official correspondence.

1844

In January, 1844, new regulations were issued "forbidding "any regimental record or device being placed on the Queen's "colour, other than the number of the regiment, in gold "characters, surmounted by the Imperial crown"; from which, it is clear, that the colours shown in Cannon's plate were intended to represent those of the latest regulation pattern up to 1848.

1854

In the Inspection Report, dated 10th October, 1853, the old colours of 1830 were reported on as "not according to "present regulations; new ones have been applied for."

The next issue of colours to the regiment was accordingly made early in 1854, prior to embarking for Gibraltar, and, up to about this period, it would appear that it was not customary in the army for colours to be formally presented, but that they were received and accounted for merely as ordinary stores.

On the regiment embarking for the Mediterranean in April, 1854, the old colours were handed over to the safe keeping of the regimental agents, by whom they were carefully preserved until 1876. They were then forwarded to Colonel Brice, commanding the 2nd Battalion, at Templemore, and a meeting of the officers was held, when it was decided to ask the Colonel of the regiment (Major-General W. Faber, C.B.) to accept charge of them, to which he gladly assented. After his death, they were eventually handed over to an escort at Leicester for removal to St. Martin's Church, where they were deposited on the 16th July, 1891, together with the first set of colours of the 2nd Battalion, which had been presented on its formation in 1859.

1858-59

In 1858, a Royal Warrant was issued altering the colours; they were to be only 3 feet 9 inches flying, and 3 feet deep, ornamented with gold and crimson fringe for the Queen's, and white and gold for the regimental colour; the poles to be surmounted with the crest of England instead of the ornamental spear-head; cords and tassels 3 feet long of crimson and gold.

The first "presentation" of regimental colours seems to have been on the formation of the 2nd Battalion in 1858, the colours having been presented by Lady Vivian on the 3rd February, 1859.

1879

The next set of colours was received by the 1st Battalion early in 1879, and awaited (at the Service depot at Rawal Pindi), the return of the battalion from active service in Afghanistan. These colours were never formally presented.

1881

With the introduction of the Territorial system, on the 1st July, there disappeared from the regimental colour the small Union in the dexter canton, which had been authorised for 130 years, it having been first approved in the Royal Warrant of the 1st July, 1751. In lieu of it, the following was directed in the "Queen's Regulations," dated 1st July, 1881. "In regiments which are faced with white, the Second "Colour is to be the red Cross of St. George, in a white field, "with the Territorial designation, and the title displayed within "the Union wreath of roses, thistles, and shamrocks, and "ensigned with the Imperial crown."

1882

On the 4th May, 1882, a letter was received from the Horse Guards stating that Her Majesty Queen Victoria had been graciously pleased to command that the victory of "Louisburg" be inscribed on the colours in addition to the present achievements of the regiment.

1885

At the second issue of colours to the 2nd Battalion (since its formation in 1858), which took place on the 4th December, 1885, it was perceived that the Royal Tiger was, for the first time, depicted on the regimental colour as a tiger "proper."

1906-10

The next set of colours issued to the 1st Battalion were received in India in 1906, when it was found that the Royal Tiger on the regimental colour was again shown as "proper." These colours were not taken into use until July, 1908. Application had been first made by Colonel G. H. P. Burne, commanding 1st Battalion, through Major-General Utterson, C.B. (Colonel of the regiment), to the Army Council, for the tiger on the colours to be reinstated green as formerly.

On the 12th February, 1908, a reply was received, stating that His Majesty the King had been graciously pleased to approve of the Royal Tiger, emblazoned on the colours of the Leicestershire Regiment, being green with yellow stripes.*

In the Clothing Regulations for 1909, Part I., Appendix 11, para. 11, it was laid down that: " In those regiments where " the number of actions exceeds nine, laurel branches are to " be introduced, and the scrolls bearing the names of the " actions entwined thereon."

In Army Orders for November, 1909, it was announced that the King had been graciously pleased to approve of the award of the following honorary distinctions, to be borne on the colours of the Leicestershire Regiment, viz.: " Martinique, " 1762 " ; " Havannah."

In the Army Orders for March, 1910, it was announced that the King was graciously pleased to approve of the Leicestershire Regiment being permitted to bear the honorary distinction, " Namur, 1695," on its colours, in recognition of services rendered during the siege and capture of that town.

The height of the staves, with ornaments, by present regulation, is 8 feet 7½ inches, and the dimensions of colours have not been altered since the Royal Warrant of 1859.

By Army Order No. 251, of 1st October, 1910, His Majesty the King directed that Battle Honours and honorary distinctions borne on the colours and appointments of regular battalions, shall in future be borne in the same way by the Special Reserve battalions affiliated to them.

1912

A statement appeared this year in February number of the Regimental Journal, to the effect that an old King's Colour of the 17th (prior to the Union in 1800) had been discovered in the village church of Rockbourne, Hampshire, which had been given to it by the owner of West Park, formerly the home of General Sir Eyre Coote, K.B., K.G. (who in 1799 had been posted as Commandant of the 2nd Battalion 17th, on its formation, to take part in the Helder Expedition), the assumption being, that, on the disbanding of the battalion in 1802, this Colour had been secured by its colonel.

* " Yellow " should apparently have read " gold."

Enquiry, however, at the Heralds' College, has elicited the fact that there are no particulars in possession as to any issue of colours to the regiment in 1799 (see Page 261), from which it would appear that it might have been one of the Colours washed up from the shipwreck mentioned under the year 1802 in this chapter.

APPENDIX.

---

1. Succession List of Colonels.
2. Biographies of the Colonels.
3. "The Truth about the Siege of Londonderry, in 1689." (A vindication of Colonel Solomon Richards.)
4. Account of the Battle of Sheriffmuir, in 1715, by Major-General Wightman, Colonel of the regiment.
5. Regimental Plate, Pictures, &c.
6. Claims for compensation for losses sustained by the 17th Regiment in the wreck of the transport "Hannah" in 1840.
7. The Regimental Chapel, St. Martin's Church, Leicester.
8. Three Presentation Cases.
9. Regimental Music.

SUCCESSION OF COLONELS

OF THE

17TH REGIMENT OF FOOT.

THE LEICESTERSHIRE REGIMENT.

Officers' Shako Plates (*exact sizes*).

1855. 1861.

General Sir F. A. Wetherall, G.C.H.

(From an oil painting.)

BIOGRAPHIES OF COLONELS
OF
THE 17TH
OR THE
LEICESTERSHIRE REGIMENT OF FOOT.

SOLOMON RICHARDS.

Appointed 27th September, 1688.

Solomon Richards had been Governor of Wexford in* Cromwell's time, and served on the Continent in the reign of King Charles II. In the autumn of 1688 he was nominated by King James II. to raise a corps of pikemen and musketeers, now the 17th Regiment of Foot, of which he was appointed colonel on the 27th of September, 1688. At the Revolution he transferred his services to the Prince of Orange, afterwards King William III., who sent him, with his regiment, to the relief of Londonderry, under the direction of Colonel Cunningham (9th Regiment), commanding troops. They returned to England, at the suggestion of the Governor of Londonderry, who stated the place could not be defended against the army advancing to attack it, and King William, disapproving of his conduct, deprived him of his commission. He was nevertheless honoured at death, by finding a resting place in the North Cloister of Westminster Abbey, 6th October, 1691.†

SIR GEORGE ST. GEORGE.

Appointed 1st May, 1689.

This officer was the second son of Sir George St. George, Knight, of Carrickdrumruck, County Leitrim. Previous to his appointment to the colonelcy of the 17th Regiment, he served in the Earl of Ardglass's Regiment of Horse in several

* Swift's Pamphlet, 1731.
† Dalton.

campaigns in Flanders, and, with the 17th, in the campaign of 1694, in Flanders. In 1695, he exchanged with Colonel Courthorpe to a newly-raised regiment, which was disbanded in 1798. In the "Clarendon State Papers" (Vol. I., page 95), he is thus referred to by the Earl of Clarendon: "Has served "ever since the King's restoration; is known to be a brave "man and as good an officer as any in the army, by all who "know Ireland."

A War Office letter to the Duke of Bolton, Governor of Ireland (entered at the Signet Office, 12th March, 1718), shows him raised to the peerage as "Lord George St. George," and appointed Governor of Galway, at £200 a year, from the 24th August, 1717.*

John Courthorpe.

Appointed 1st *May*, 1695.

John Courthorpe, only son and heir of Sir Peter Courthorpe, of Court Town, County Cork, entered the Army in the time of King Charles II., and afterwards commanded a company of foot. He was appointed colonel of one of the regiments raised for the reduction of Ireland in 1689.

In 1692, he fought with King William's army at the battle of Steenkirk, where he was wounded and taken prisoner,† and in 1695 exchanged to the 17th Foot. He served in the Netherlands under King William III., and was killed at the head of his regiment when storming the breach of Terra Nova at the Castle of Namur, on the 30th of August, 1695.

Sir Matthew Bridges.

Appointed 1st *September*, 1695.

After a progressive service in the subordinate commissions, this officer was appointed lieut.-colonel in the 17th Regiment. On December 4th, 1691, he was appointed Governor of Londonderry and Culmore by Queen Mary II. He produced Her Majesty's commission, and required from the Irish Society the usual salary of £200 a year, payable to the Governor of Culmore Fort in future.‡

* W.O. 8, Vol. I., p. 102.
† Dalton, Vol. III., p. 42.
‡ John Mackenzie.

He distinguished himself at the storming of the breach of Terra Nova at the Castle of Namur, on the 30th of August, 1695, when he was wounded. King William III. conferred the colonelcy of the regiment upon him, and he served under His Majesty until the Treaty of Ryswick in 1697; and subsequently commanded his regiment in Ireland.

From 1702 to 1706 he was again appointed (at 10/- a day), Governor of Londonderry. The date of his decease has not been ascertained.*

Holcroft Blood.

Appointed 26th August, 1703.

This officer was the son of the celebrated Colonel Thomas Blood, who made a desperate effort to carry off the crown from the Tower of London in the reign of King Charles II., for which the Colonel was afterwards pardoned, in consequence of his previous services in the Royal cause. Holcroft Blood served on board the fleet, in the war with Holland in 1672 and 1673, and subsequently entered the French Army as cadet in the guards of Louis XIV., where he attained great proficiency in the study of fortification. At the Revolution in 1688 he returned to England and was appointed to a commission in Colonel Seymour's Regiment, in which corps he rose to the rank of major. He served in Ireland, where he was employed as an engineer, and evinced ability at the sieges of Athlone and Limerick; was wounded in action at Cavan (March, 1690), and was present at the taking of Cork and Kinsale. He also distinguished himself as chief engineer at the siege of Namur, in 1695, and accompanied the 17th to Holland in 1701. Was promoted to the lieut.-colonelcy in July, 1702, and in the same year served as a principal engineer at the sieges of Venloo and Ruremonde, where he displayed great ability. He particularly distinguished himself at the storming of Fort St. Michael, at Venloo, where "he showed the part of a brave officer, charging "with the men sword in hand, and killing an officer of the "enemy's grenadiers, who made a vigorous opposition with "his party."†

* Dalton.

† Boyer's "Annals of Queen Anne."

The talents and bravery of Colonel Blood procured him the favour of the great Duke of Marlborough, who obtained for him the colonelcy of the 17th Regiment and the command of the British artillery on foreign service on the Continent, with the rank of brigadier-general.

The "Blenheim Roll" shows Colonel Blood as Colonel of a "Train of Artillery," raised by Royal Warrant, dated March 14th, 1702.

At the memorable battle of Blenheim, in 1704, when detached from the regiment, Brigadier-General Blood highly distinguished himself, and by bringing nine field-pieces, loaded with cartridge shot, into action at a critical moment, greatly contributed to the gaining of that splendid victory; by a General Order issued in the evening of that day, all the trophies captured were placed under his care.

A hundred cannon were taken from the enemy, and for his services at Blenheim he received a bounty of £75.* Towards the close of the campaign he accompanied the Duke of Marlborough to the Moselle, and was engaged in the capture of several places in that quarter. He continued in the command of the British artillery on the Continent, and his services were associated with the forcing of the French lines at Helixem and Neer Hespen in 1705, the splendid victory at Ramilies in 1706, and the siege of Menin.*

On the 1st of January, 1707, he was promoted to the rank of major-general. He died at Brussels on the 20th August, 1707.

This officer's plans for the defence of Bruges "against the "insults of the French" in 1696 are still in existence.†

Joseph Wightman.

Appointed 20*th August*, 1707.

Joseph Wightman commenced his career as ensign in the 1st Foot Guards on the 29th December, 1690, in which he became a captain and lieut.-colonel in December, 1696, and on transfer to the 17th, he served with it in the Netherlands

* Dalton.

† King William III.'s Chest (formerly a sealed bag), 1696–97, No. 16, Public Record Office.

under King William III., and accompanying the regiment to Holland in 1701, served through the campaigns of 1702 and 1703 under John Duke of Marlborough. He was promoted to the lieut.-colonelcy of the regiment in 1702, and to the rank of colonel in the army on the 26th August, 1703. He served in Flanders, Portugal, and Spain, under the Earl of Galway; was nominated brigadier-general on the 1st January, 1707, and to the colonelcy of the 17th Regiment in August following; on the 1st January, 1710, he was promoted to the rank of major-general, and, in the absence of the Duke of Argyll, became Commander-in-Chief in Scotland on the 13th June, 1712, and commanded the centre division of the Royal Army at the battle of Sheriffmuir in November, 1715. He wrote an account of the battle, which was published at the time. (See p. 304.)

In 1719 he commanded the King's troops at the battle of Glenshiel, when he forced the Highlanders to disperse, and the Spanish troops to surrender prisoners of war. His services were rewarded with the government of Kinsale, from the 25th March, 1720, and he was appointed a major-general on the Irish establishment from the 14th August, 1721.* He died suddenly at Bath, of a fit of apoplexy, on the 25th September, 1722,† which is at variance with the Scotch account of his having been present, as a spectator, at the battle of Preston Pans in 1745. (" Culloden Papers.")

THOMAS FERRERS.

Appointed 28th September, 1722.

This officer served under the celebrated John Duke of Marlborough, and was promoted to captain and lieut.-colonel in the Foot Guards; in May, 1705, he was advanced to the rank of colonel, and in 1710 to that of brigadier-general. Being conspicuous for loyalty at a period when Jacobite principles were prevalent in the kingdom, he was commissioned to raise a regiment of dragoons, which was disbanded in 1718; and in the following year he was appointed colonel of the 39th Foot, from which he was transferred, in September, 1722, to the 17th Regiment. He died on the 26th October following.

* W.O. 8, Book 2, p. 4.
† " Irish Martial Affairs," Book 17, p. 41.

James Tyrrell.

Appointed 7th November, 1722.

James Tyrrell was appointed ensign in a regiment of foot on the 6th of February, 1694, and he served under King William III. in the Netherlands. He distinguished himself in the wars of Queen Anne; and was promoted to the colonelcy of a newly-raised regiment of foot in April, 1709. At the peace of Utrecht his regiment was disbanded; and in 1715 he raised a regiment of dragoons for the service of King George I., which was disbanded in November, 1718: in 1722 His Majesty gave him the colonelcy of the 17th Regiment. He was promoted to the rank of brigadier-general in 1727, and placed on the Irish establishment at £1 a day, from the 28th March, 1732; promoted major-general in 1735 and lieut.-general in 1739. In February, 1740, he was governor of Tilbury Fort, and on the 14th August, same year, was appointed to a command in Ireland at £485 per annum.* He died on the 1st August, 1742.

John Wynyard.

Appointed 31st August, 1742.

John Wynyard was many years an officer of the 17th Regiment of Foot, with which corps he served in the Peninsula in the War of the Spanish Succession, and in Scotland during the Earl of Mar's rebellion. On the 10th of July, 1718, he was promoted to the lieut.-colonelcy of the regiment, and his zealous attention to all the duties of his situation was rewarded, in November, 1739, with the colonelcy of the 4th Regiment of Marines, which was then newly raised, from which he was removed, in 1742, to the 17th Regiment, which corps he had commanded many years with reputation. He died in 1752.

Edward Richbell.

Appointed 14th August, 1752.

This officer entered the army in the reign of Queen Anne, and served with reputation under the Duke of Marlborough. He evinced a constant attention to the

* W.O. 26, Book 19, p. 286.

duties of his profession, and was promoted, on the 18th May, 1722, to the lieut.-colonelcy of the 37th Regiment. He distinguished himself in the War of the Austrian Succession, and was promoted to the colonelcy of the 39th Regiment on the 14th June, 1743. In 1746 he commanded a brigade under Lieut.-General St. Clair, in the expedition against Port L'Orient; and in 1752 he was removed to the 17th Regiment. He died in 1757.

JOHN FORBES.

Appointed 25th February, 1757.

John Forbes obtained a commission in the army on the 10th of April, 1710; after a progressive service in the subordinate commissions, and distinguishing himself in the War of the Austrian Succession, he was promoted to the lieut.-colonelcy of the Scots Greys on the 29th of November, 1750: in 1757 he was advanced to the colonelcy of the 17th Regiment. He was nominated Adjutant-General to the expedition against Louisburg in 1757; and afterwards appointed Commander-in-Chief of the troops in the southern provinces of North America, with the rank of brigadier-general. He died on the 11th of March, 1759.

THE HONOURABLE ROBERT MONCKTON.

Appointed 24th October, 1759.

The Honourable Robert Monckton, son of John, first Viscount Galway, served in the army in the War of the Austrian Succession; and in February, 1751, he was promoted to the lieut.-colonelcy of the 47th Regiment: in 1757 he was nominated colonel-commandant of the Second Battalion of the 60th Regiment. He commanded a brigade, under Major-General James Wolfe, in the expedition against Quebec, and evinced great gallantry and ability on several occasions; he was shot through the lungs at the battle on the Heights of Abraham on the 13th September, but recovered of his wound and was nominated Lieut.-Governor of Annapolis Royal and colonel of the 17th Regiment. In 1761 he was appointed Governor and Commander-in-Chief of the province of New York, and promoted to the rank of major-general on the 20th February. Soon afterwards he was selected to command the land forces of an expedition against the French island of Martinique, which he captured, after overcoming numerous difficulties, early in 1762. He was nominated Governor of Borwick and Holy Island, and afterwards of Portsmouth, which place he represented in Parliament several years. He was promoted to the rank of lieut.-general in 1770. His decease occurred on the 21st of May, 1782.

GEORGE MORRISON.

Appointed 29th May, 1782.

This officer served many years on the staff of the army; he was advanced to the rank of lieut.-colonel in 1761, at which period he held the appointment of deputy-quartermaster-general; and in 1763 he was placed at the head of that department. He was promoted to the rank of colonel in 1772, and to that of major-general in 1777; in 1779 he was appointed colonel of the 75th Regiment (afterwards disbanded), from which he was removed, in 1782, to the 17th, and also promoted to the rank of lieut.-general. He was appointed to the 4th Regiment of Foot in 1792; and promoted to the rank of general in 1796. He died in 1799.

GEORGE GARTH.

Appointed 8th August, 1792.

This officer served thirty-seven years in the 1st Regiment of Foot Guards, in which corps he was appointed ensign and lieutenant at the commencement of hostilities with France in 1755. In 1758 he obtained the rank of lieutenant and captain, and he afterwards served in Germany under Prince Ferdinand of Brunswick: on the 6th of February, 1772, he was promoted to the rank of captain and lieut.-colonel. When the American War commenced, his services were extended to that country, where the Foot Guards had opportunities of distinguishing themselves. He was promoted to the rank of colonel in 1779; was nominated major in his regiment in March, 1782, and advanced to the rank of major-general in November following: in 1789 he was appointed lieut.-colonel in his regiment. King George III. was pleased to confer on Major-General Garth the colonelcy of the 17th Regiment in 1792; also to promote him to the rank of lieut.-general in 1796, and to that of general in 1801. General Garth was subsequently appointed lieut.-governor of Placentia. He died in 1819.

JOSIAH CHAMPAGNÉ, G.C.H.

Appointed 14th June, 1819.

On the 28th of January, 1775, Josiah Champagné was appointed ensign in the 31st Foot, and embarking with his regiment, in March, 1776, for the relief of Quebec, then besieged by the Americans, he arrived in Canada in May, and took part in the operations by which the troops of the United States were forced to quit the British provinces. He remained on active service in Canada during the remainder of the American War, was promoted to a lieutenancy in his regiment in July, 1777, and, returning to England at the peace in 1782, was nominated captain in the 99th Foot (afterwards disbanded) in 1783, and removed to the 3rd Foot in March, 1784. He joined the Buffs at Jamaica in May of the same year; and in 1789, when the Nootka Sound question threatened to involve Great Britain and Spain in war, he embarked with a detachment of his regiment on board the fleet; he returned to England soon afterwards. He again embarked for the West Indies, with his regiment, in 1793—the Buffs forming part of the expedition under Lieut.-General Sir Charles Grey; but their destination was afterwards changed to Ostend; and they subsequently joined the armament under Major-General the Earl of Moira, prepared to aid the French loyalists. In the same year Captain Champagné was promoted to a majority in the 80th Foot, and afterwards to a lieut.-colonelcy in the same corps. In 1794 he again proceeded to the Continent, and, after serving in the retreat through Holland, returned to England. He embarked for the coast of France in 1795, and served with the expedition under Major-General Doyle which took possession of Isle de Dieu. In 1796 he proceeded with his regiment to the Cape of Good Hope, and towards the close of the same year sailed to the East Indies. He was promoted to the rank of colonel in 1797; and in 1800 was nominated to command an expedition against Batavia, with the rank of brigadier-general, but this enterprise was countermanded; on the 25th April, 1801, was appointed colonel of a regiment of infantry, which later became the 1st Ceylon Regiment, in 1818 a regiment of light infantry, and in 1822 "The Ceylon Rifles" (disbanded in 1873). Brigadier-General Champagné was afterwards named second in command of the army which proceeded from India to Egypt in 1801. He returned to England in 1803; and in September of that year was promoted to the rank of major-general. On the 22nd of February, 1810, he was transferred from the colonelcy of the 1st Ceylon Regiment to that of the 41st Foot; and in July following promoted to the rank of lieut.-general. In 1819 he was transferred to the 17th Regiment. He was honoured with the dignity of Knight Grand Cross of the Royal Hanoverian Guelphic Order; and was advanced to the rank of general in 1821. He died on the 31st of January, 1840.

Sir Frederick Augustus Wetherall, G.C.H.

Appointed 17th February, 1840.

This officer entered the service in August, 1775, as ensign in the 17th Foot. He embarked at Cork with the regiment in September following for Boston, North America, where he remained during the siege, and accompanied his corps at the evacuation to Halifax in March, 1776. In June following he proceeded with the army under the command of Sir William Howe to Staten Island, preparatory to the attack of New York. In August, 1776, he received a lieutenancy; in which rank he served five years, and was constantly employed in North America and Europe. He was present at the battles of Brooklyn, Whiteplains, Fort Washington, Princetown, Brandywine, Germantown and Monmouth, exclusive of several affairs of posts, in North America. He was embarked and did duty as a captain of marines on board His Majesty's Ship "Alfred," and was in the battles of Cape Finisterre and St. Vincent, under Sir George Rodney, previous to the relief of Gibraltar. On the 17th May, 1781, he raised an independent company, which was embodied in the 104th Regiment, and was employed on the island of Guernsey. On 16th April,

GUARD OF HONOUR TO THE AMIR YAKOOB KHAN, GUNDAMUK, MAY, 1879.

(By the courtesy of the India Office.)

OFFICER'S SHAKO PLATE (*exact size*). 1871—79.

OFFICER'S HELMET PLATE (*reduced in size*)
1879 TO 30TH JUNE, 1881.

1783, he exchanged into the 11th Regiment, and proceeded to Gibraltar, where he did duty six years. In 1790, he attended the Duke of Kent to Quebec, and accompanied His Royal Highness as aide-de-camp to the West Indies in 1794; he was at the taking of Martinique, where he received two wounds. On 1st March, 1794, he was appointed major in the 11th Foot, and employed as deputy-adjutant-general to the forces in Nova Scotia, under the command of the Duke of Kent, to which situation he was appointed on 23rd August, 1794. On 20th May, 1795, he was appointed lieut.-colonel in Keppel's* Regiment, and employed at St. Domingo under the command of Lieut.-General Sir Adam Williamson and Major-General Forbes; he was entrusted by the latter officer with despatches for Sir Ralph Abercromby at Barbados, and on the passage was taken by a French frigate, and wounded in the action; he remained at Guadaloupe a prisoner of war upwards of nine months, and when exchanged was appointed adjutant-general to the forces under the command of the Duke of Kent in North America. On the 3rd August, 1796, he was removed to the lieut.-colonelcy of the 82nd Regiment, and on 29th April, 1802, received the brevet of colonel. He afterwards raised the Nova Scotia Fencible Regiment in North America, of which he was appointed colonel on 9th July, 1803, and adjutant-general and brigadier to the forces on the Caribbee Island station in May, 1806. On the 25th of October following he was removed to the Cape of Good Hope, when he served as brigadier to the forces in that colony until 1809; he obtained the rank of major-general on 25th October of that year, and was appointed to the staff in India. On his passage from the Cape to India he was again taken prisoner in the Company's Ship "Wyndham," after a severe action, by a French squadron, in the Mozambique Channel, and carried to the Isle of France, when, after being confined two months, he was exchanged, and sailed for Calcutta. He served as second in command, under Sir Samuel Auchmuty, on the expedition against Java, which terminated in its conquest.

For his services on that occasion he had the honour to receive a medal and the thanks of both Houses of Parliament.

* One of eight regiments, raised in 1795, for service in the West Indies. This corps eventually became the 3rd West India Regiment.

His next appointment was to the command in Mysore and its dependencies, which he held until June, 1815, when he returned to England and was appointed equerry and comptroller of the household of the Duke of Kent. He was with the Duke when the latter died, and executor to his will, and afterwards served in the Duchess's household, having been made by the Duke a guardian of the Princess, who became Queen Victoria.

He received the rank of lieut.-general on 4th June, 1814. On 10th January, 1837, he was advanced to the rank of general, and His Majesty King William IV. conferred upon him the colonelcy of the 62nd Regiment. On 17th February, 1840, the Queen bestowed upon him the colonelcy of the 17th Regiment, in which he had commenced his military career. He died at Ealing on the 18th of December, 1842, having attained the advanced age of eighty-eight years.

Sir Peregrine Maitland, K.C.B.

Appointed 2nd January, 1843.

Gazetted ensign, 25th June, 1792, and promoted to captain and lieut.-colonel in the Foot Guards in 1803. He served in Flanders and was present in several actions; served also at Ostend in 1798. Was present in the actions of Lugo and Corunna, in the Peninsular War, and received a medal for the passage of the Nive. Was also at the battle of Waterloo. Sir Peregrine was advanced to the rank of colonel in January, 1812, and to the ranks of major-general and lieut.-general in 1814 and 1830 respectively. In 1843, transferred from the 76th Regiment to the colonelcy of the 17th. He died with the rank of general on the 30th May, 1854.

Thomas John Wemyss, C.B.

Appointed 31st May, 1854.

Entered the army on the 9th January, 1803. Served with the Walcheren expedition in 1809, and subsequently in the Peninsula as a brigade-major from 1810 to 1814. Was present at the battles of Fuentes D'Onor and Vittoria (brevet rank of major), at Donna Maria (severely wounded), battles of the Nivelle, Cambo, Nive, St. Pierre (wounded), Toulouse, and numerous minor affairs. Received the war medal with seven clasps. Served also against the Candians in Ceylon.

Became lieut.-colonel in 1819, and when placed on half pay the following year, was appointed assistant adjutant-general at Manchester. Promoted colonel in 1837; major-general 9th November, 1846; and lieut.-general 20th June, 1854. He died on the 19th July, 1860.

SIR RICHARD AIREY, K.C.B.

Appointed 20th July, 1860.

Obtained his first commission as ensign on the 15th March, 1821, and was advanced to the rank of lieut.-colonel in 1838, and colonel in 1851. He served throughout the Eastern campaign of 1854–55, first in command of a brigade, and afterwards, from the disembarkation in the Crimea, as quartermaster-general, and was present at the battles of Alma, Balaklava and Inkerman, the siege and fall of Sevastopol, for which (in addition to the Crimean and Turkish medals) he received the second class of the Medjidie, and was appointed a Knight Commander of the Bath, Commander of the Legion of Honour, and Commander of the First Class of the Military Order of Savoy. Promoted to major-general, 12th December, 1854; lieut.-general, 24th October, 1862; and on the 1st May, 1868, was transferred to the colonelcy of the 7th Fusiliers. He later became Lord Airey, received the Grand Cross of the Bath, and was promoted to the rank of general on the 9th April, 1871.

JOHN GRATTAN, C.B.

Appointed 1st May, 1868.

Was gazetted ensign on the 8th July, 1813, and obtaining his captaincy in 1836, was actively employed on the frontier during the rebellion in Canada in 1838. He served in China with the 18th Royal Irish; was present at the storming of the heights above Canton, and led the advance against the enemy's entrenched camp, for which service he was selected by Sir Hugh Gough as the bearer of his despatches. Was promoted to the rank of major, and appointed brigade-major of Fort St. George by Lord Hill. On his return to China on board the "Madagascar," in charge of Lord Auckland's despatches, the ship having caught fire during a gale of wind, he, with a few others, narrowly escaped the fate of fifty-seven souls, who perished on that occasion ; he fell into the hands of the Chinese and was detained 108 days in captivity, after which he was at the attack of Segoan, and commanded the 18th at Chapoo, Woosung, and Shanghai, and was present at the storming of Ching Kiang Foo and the landing before Nankin.

Served with the 18th Royal Irish in the campaign in Burmah. Was promoted colonel, 20th June, 1854 ; major-general, 10th March, 1861 ; lieut.-general, 15th September, 1870, and died on the 29th April, 1871.

William Raikes Faber, C.B.

Appointed 30th April, 1871.

Received his commission as ensign on the 10th April, 1826. Served with the 49th Regiment throughout the war in China, and was present at the first and second captures of Chusan, storm and capture of the heights above Canton, attack and capture of Amoy, and of Chinhae (mentioned in despatches and brevet of major), occupation of Ningpo and repulse of the night attack ; also the attack and capture of the enemy's entrenched camp on the heights of Segoan, of Chapoo, and Woosung, and investment of Nankin. In June, 1854, was promoted colonel in the 35th Regiment. Served in the Mutiny Campaign in 1857–58 in command of the 53rd Regiment, and was present at the battle of Cawnpore on the 6th December, 1857, and in the affair at Seraighat (mentioned in despatches). Was promoted major-general, 9th March, 1863 ; lieut.-general, 5th December, 1871 ; and general, 1st October, 1877. He died on the 24th June, 1879.

Richard W. P., Earl Howe, C.B.

Appointed 25th June, 1879.

Gazetted as ensign and lieutenant in the Grenadier Guards on the 14th July, 1838, and becoming lieutenant and captain in April, 1844, he served as aide-de-camp to Sir George Cathcart in the Kaffir War in 1852–53 (medal), for which he received the brevet rank of major. Became a captain and lieut.-colonel in June, 1854, and colonel in 1857, in which year he served at the siege of Delhi as acting quartermaster-general of the Queen's troops (C.B., medal and clasp). Was promoted major-general, 9th March, 1863; lieut.-general, 5th December, 1871; and general on the 16th March, 1880. On the 13th June, 1890, was transferred to the colonelcy of the 2nd Life Guards.

JOHN CHRISTOPHER GUISE, V.C., C.B.

Appointed 13th June, 1890.

Entered the army in 1845, and obtaining his captaincy in 1854, served with the 90th Light Infantry in the Crimea and was present at the siege of Sevastopol. He also took part in the Mutiny Campaign, including the second relief of Lucknow, where he led the attack on the Secundrabagh, and received the Victoria Cross for conspicuous gallantry in action on November 16th and 17th, 1857, no details of which are given in the official documents beyond that he was chosen by the officers of his regiment as being the most worthy and distinguished of them all, some forty in number. He attained the rank of colonel, 7th August, 1863; became major-general, 23rd March, 1869; lieut-general, 1st July, 1881; and died, much regretted, on the 5th February, 1895.

SIR JOHN ROSS, G.C.B.

Appointed 6th February, 1895.

Gazetted as second-lieutenant in the Rifle Brigade on the 14th April, 1846, and served in the Eastern Campaign, including the battles of Alma, Inkerman, and siege of Sevastopol, for which he was awarded the brevet of major and the 5th Class of the Medjidie. Served also in the Mutiny Campaign of 1857–58, including the battle of Cawnpore and operations for relief of Lucknow. Commanded the Camel Corps at the capture of Calpee, and during the subsequent campaign in Central India (mentioned in despatches, brevet of lieut.-colonel, C.B.). Served with the Rifle Brigade in the campaign of the North-west frontier of India, 1863–64. Promoted major-general on the 1st March, 1870, and commanded the Bengal troops during the operations in the Malay Peninsula in 1875–76 (mentioned in Government of India General Orders). Commanded the Indian expeditionary force, sent to the

Mediterranean in 1878. Served in the Afghan War of 1878–80, and commanded the 2nd Division Cabul Field Force, which defeated the enemy at Shekabad (received the thanks of the Governor-General in Council, and of the Commander-in-Chief in India). Accompanied Sir Frederick Roberts on the march to Candahar, and was present at the battle when in command of the infantry division. (Mentioned in despatches, K.C.B., and received the thanks of both Houses of Parliament.) Promoted lieut.-general, 12th January, 1886, and was later appointed a Knight Grand Cross of the Bath. Promoted to general on the 1st April, 1891, and died on the 28th July, 1903.

George Tito Brice, C.B.

Appointed 29th July, 1903.

Received his first commission as ensign in the 17th Foot on the 22nd December, 1848, and was adjutant from November, 1850, to July, 1852. Obtaining his captaincy in 1854, he served in the Crimean Campaign of 1855, including the siege and fall of Sevastopol and the assaults on the Redan of the 18th June and 8th September, and was present at the bombardment of the fortress of Kinbourn. Mentioned in despatches, and (in addition to the Crimean and Turkish medals) received the 5th Class of the Medjidie and a brevet majority. Was promoted lieut.-colonel in July, 1865, and colonel in March, 1874. From 1879 to 1884, he was in command of a brigade of the Bombay Army, and promoted major-general on the 1st April, 1885, being placed on the retired list in 1889. On the occasion of the Jubilee of the Crimean War he was made Companion of the Bath, on the King's birthday in 1905. He died, aged 74, on the 22nd September, 1905.

Archibald Hammond Utterson, C.B.

Appointed 23rd September, 1905.

Served for 33 years with the regiment, and was gazetted to it on the 25th August, 1854. Served in the Crimean Campaign, including the siege and fall of Sevastopol, assault on the Redan, and bombardment and surrender of Kinbourn. (Medal with clasp and Turkish medal.)

Was promoted brevet lieut.-colonel on the 1st October, 1877, and served in 1878 in the Afghan War. Was present at the capture of Ali Masjid, the action of Deh Suruk, both expeditions into the Bazar Valley, and other minor affairs. Mentioned in despatches (" London Gazette," 7th November, 1879) and medal and clasp. Appointed to the command of the First Battalion Leicestershire Regiment in 1884, to the command of the 17th Regimental District in 1888, and after promotion to major-general on the 28th January, 1891, was selected for the command of a brigade at Aldershot in January the following year, which he held until retiring on half pay in January, 1895. In 1893, was nominated a Companion of the Bath, in the following year was granted the Reward for Distinguished and Meritorious Service, and was placed on the retired list on the 10th March, 1898.

THE TRUTH
ABOUT
THE SIEGE OF LONDONDERRY IN 1689.

A VINDICATION
OF
COLONEL SOLOMON RICHARDS.

A review of the circumstances attending the despatch to Ireland of Colonel Cunningham's (9th) and Colonel Richards' (17th) Regiments, in April, 1689, and their early return to England.

Compiled from old works of the above date, from which the immoderate use of capital letters and occasional bad spelling, &c., have been faithfully reproduced.

FROM KING WILLIAM III.

INSTRUCTIONS FOR OUR TRUSTY AND WELL-BELOVED ROBERT LUNDY, ESQ., GOVERNOR OF OUR CITY, AND GARRISON OF LONDONDERRY, IN OUR KINGDOM OF IRELAND.

WHEREAS, We have thought fit to send two of Our regiments of foot under the command of Colonel Cunningham, and Colonel Solomon Richards, for the relief of Our City of Londonderry ; we do hereby authorise and empower you, to admit the said regiments into Our said City, and to give such orders, concerning their quarters, duty, and service, during their stay in those parts, as you shall think fit for the security of the said city, and country thereabouts. And, whereas, we are sending to Our said city of Londonderry, further succours of money, men, arms and provisions of War ; we do expect, from your courage, prudence, and conduct, that, in the meantime, you make the best defence you can against all persons that shall attempt to besiege the said city, or to annoy Our Protestant subjects with the same, or within the neighbouring parts ; and, that you hinder the enemy from possessing themselves of any passes near, or leading to the said city ; giving all aid, and assistance you may, with safety, to such as shall desire it, and receiving into the said town, such Protestant officers and men, able and fit to bear arms, as you may confide in, whom you are to form into companies, and to cause to be well exercised and disciplined. Taking care withal, that you do not take in more unuseful people, women and children, into the said city, than there shall be a provision sufficient to maintain besides the garrison. You are to give us an account as soon as may be, and, so, from time to time, of the condition of our city of Londonderry, the fortifications, number, quality and affections of the people, soldiers, and others therein, or in the country thereabouts ; and, what quality of provisions of all sorts for horse, foot, and dragoons, shall, or may be brought up, or secured in those parts for Our Service, without the necessity of bringing the same from England, upon sending of more forces thither.

Lastly, We do recommend unto you, THAT you entertain good correspondence and friendship with the officers of the said regiments, and more especially with the respective colonels of the same; not doubting, but by your joint councels, and by your known courage, as well as your affection to the Protestant religion, which, we shall not fail to reward with Our Royal favour and bounty, the said city will continue under our obedience, until upon the arrival of an Army we are sending from England, all things shall be in such a posture, as that we may there, with the blessing of God, restore, in a short time, Our Kingdom of Ireland, to its former peace and tranquility.

Given at Our Court at Whitehall, the 12th Day of March, 1688–89, in the first year of Our Reign.

By His Majesty's Command.

N.B.—It should be mentioned at the outset, that, by date of commission (27th September, 1688), Colonel Richards was senior to Colonel Cunningham, whose commission bore date, 31st December, 1688, but, for some unknown reason, the supreme command of the troops (who embarked for Londonderry in April, 1689), was given to Colonel Cunningham, as the Officer Commanding *the senior regiment.*

ORDERS AND INSTRUCTIONS, FOR OUR TRUSTY AND WELL-BELOVED JOHN CUNNINGHAM ESQRE, COLONEL OF ONE OF OUR REGIMENTS OF FOOT, AND UPON HIS DEATH OR ABSENCE, TO COL. SOLOMON RICHARDS, OR TO THE OFFICER IN CHIEF WITH THE REGIMENTS, WHEREOF THEY ARE COLONELS.

WILLIAM R.

You are, without delay, to repair to the Quarters of the Regiment under your Command, and take care that it be in a readiness to March to Liverpool, at such a time as you shall Appoint. Whereupon, you are to go to Liverpool, & to Enquire what Ships there are in that Port, appointed to carry over the two Regiments, whereof you and Solomon Richards are Colonels, to the Town of Londonderry; and, whether the Frigat, ordered for their convoy be arrived there; and, as soon as the said Ships & Frigat shall be in a readiness to sail, and fitted with all Provisions, necessary for the sustenance of the said Regiments in their Passage to the said Town, and for their return from thence, if there be occasion. You are to cause Colonel Richards' Regiment to go on Board, and, at the same time, to Order the Regiment, whereof you are Colonel, to March to Liverpool, and to Embarque with all speed.

And, whereas, We have Ordered one thousand Arms to be carried to Liverpool, you are to cause such a number of the said Arms as shall be wanting in the said Regiments, to be delivered unto them, and the residue of the said Arms and Stores now there, to be put on Shipboard, and carried to Londonderry, to be employed for Our Service, as the Governour of the said Town and you shall think fit.

And, We, having also directed the sum of Two thousand pounds sterling to be paid unto you at Chester, by Matthew Anderton, Esq., Collector of Our Customs there, you are hereby authorised and required to receive the same, and to dispose of the said sum towards the necessary subsistence of the said regiments, and for the defence of the place, in repairing and providing what shall be defective therein, and to such other uses, as you, with the Governour of the said City, with whom you are to entertain a good Correspondence and Friendship, as you shall find necessary for Our Service ; all of which Expences you are to give Us an account by the first opportunity. When the Particulars necessary for the Voyage shall be fully complied with, you are then, Wind and Weather permitting, with the Regiments under your Command, to make the best of your way to Londonderry, and being arrived there, or near that place, you are to make enquiry, whether the said City be yet in the hands of the Protestants ? and whether, you may, with safety, put Our said Regiments into the same ? and in that case, you are immediately to acquaint Lieutenant Colonel Robert Lundy, Our Governour thereof, or the Commander in Chief for the time being, with Our Care in sending those Regiments and Stores ; and, for the further relief of Our Protestant Subjects in those parts, and delivering him Our Letters and Orders to him directed, you are to Land the said Regiments and Stores, and to take care that they be well Quartered and disposed of in the said city, following such Directions as you shall receive during your stay there, from Our said Governour, Lieutenant Colonel Robert Lundy, in all things relating to Our Service.

You are to assure the Governour and Inhabitants of Londonderry, of further and greater Succours of Men, Arms, Money and provisions of War coming speedily from England for their relief, and the security of those parts, and in the mean time, You are to make the best defence you can against all persons that shall attempt to Besiege the said City, or to annoy Our Protestant Subjects within the same.

You are to give Us an account soon after your Arrival, (and so, from time to time), of the condition of the place, the Fortifications, number, quality and affection of the People, Soldiers, and others therein, or in the Country thereabouts, and what quantity of Provisions of all sorts for Horse and Foot, and Dragoons, shall, or may be bought up or secured in those parts for Our Service, without the necessity of bringing any from England, upon sending more Forces thither. You are to inform Us, whether Captain James Hamilton be Arrived at Londonderry, and how he has disposed of the Money and Stores, committed to his Charge, and, in general, you are to return Us an account of everything, which you, in your discretion, shall think requisite for Our Service.

In case you shall find it unsafe to Land the said Regiments at or near Londonderry, so as to put them into the Town, which you are to endeavour, by all reasonable and prudent means, you are not to expose them to extraordinary hazard in so doing, but to take care that they be carried in the same Ships, and under the same Convoy, with the same Armes, Stores, Money and Provisions above mentioned, to Carrickfergus, and to endeavour the Landing of them there, if the same may be done with safety, or otherwise to Strangford, at both or either of which places, you are to use the same caution, and, to follow as near as may be, the like directions as are now given you in relation to Londonderry, but, in case, you do not find it for Our Service, to Land the said Regiments at any of the said Places, you are then to take care, that they be brought back to the Port of Liverpool, giving Us speedy notice for Our further Orders.

Given at Our Court at Whitehall, the Twelfth of March, 1688–9, in the first year of Our Reign.

By His Majesties Command,

(Signed) Shrewsbury.

ADDITIONAL INSTRUCTIONS, FOR OUR TRUSTY AND WELL-BELOVED COLONEL JOHN CUNNINGHAM, OR THE OFFICER IN CHIEF, WITH OUR TWO REGIMENTS OF FOOT, WHEREOF HE AND COLONEL RICHARDS ARE COLONELS.

Whereas, We have ordered £2,000 sterling to be paid unto you by several Bills of Exchange, over and above the £2,000 you shall receive from Our Collector in the Port of Chester ; You are accordingly to receive the same, and upon your Arrival, at Our City of Londonderry, to pay £500 thereof to Our Trusty and Well beloved Robert Lundy Esquire, Governour thereof, as of Our Royal Bounty, in part of the reward, We intend him for his faithful services ; and the residue of the said £2,000, you are to apply towards the defraying the contingent charges, which Our said Governour, yourself, and Colonel Richards shall find requisite for the security of that Garrison, or of such other place, where our said Regiments shall Arrive, or be put on Shoar. Provided, always, that you do not in any manner, put off or delay the departure of Our said two Regiments from Liverpool to Londonderry, in case the said sum be not immediately paid unto you, by the respective Persons, from whom it is to be received.

Given at Our Court, at Whitehall, the 14th of March, 1688/9, in the first year of Our Reign.

By His Majesties Command,

(Signed) Shrewsbury.

April 15th, 1689. Colonel Cunningham and Colonel Richards came into the Lough from England with two Regiments, and other Necessaries for Supply of Derry.

They had particular instructions to receive, from time to time, such orders as Colonel Lundy should give them, in all things relating to His Majesty's service, pursuant to which, Colonel Cunningham sent three messages to him.

By the first (from Green Castle, about 10 in the morning), he acquainted him of his coming, and desired his orders, about landing the two regiments on board; by the second message (from Red Castle, about 2 in the afternoon), having some information about Col. Lundy having gone out to fight the enemy (King James's army) at Claudy, he wrote the following letter.*

LETTER FROM COLONEL CUNNINGHAM TO LT.-COL. LUNDY, GOVERNOR OF LONDONDERRY.

From on Board the "Swallow," near Redcastle, at two Afternoon, the 15th of April, 1689.

Sir,

Hearing you have taken the Field, in Order to Fight the enemy, I have thought it necessary for His Majesty's Service, to let you know, there are two well-disciplined Regiments here on Board, that may joyn you in two days at farthest. I am

* Derriana, p. 31.

sure, they will be of great Use in any Occasion, but, especially for the Encouragement of Raw Men, as I judge most of yours are:

Therefore, it is my Opinion, that you only stop the Passage of the Enemy at the Foords of Finn, till I can joyn you, and, afterwards, if giving Battel be necessary, you will be in a much better posture for it than before. I must ask your Pardon, if I am too free in my Advice; according to the Remote Prospect I have of things, this seems most Reasonable to me, but as His Majesty has left the whole Direction of Matters to you, so you shall find that no man living shall more cheerfully Obey you than,

Your most humble Servant,

John Cunningham.

Colonel Cunningham, having no answer to either of his letters, sent a third messenger from Culmore Castle, about 9 at night, to desire Colonel Lundy's orders, which he was ready to execute, but he received no answer from Governor Lundy until that evening.*

LETTERS FROM COLONEL LUNDY IN REPLY TO COLONEL CUNNINGHAM.

(1)

April 15th, Londonderry.

Sir,

I am come back much sooner than I expected when I went forth; for having numbers placed on Finn Water, as I went to a pass, where a few might oppose a greater number than came to the place, I found them on the run before the enemy, who pursued with great vigour, and I fear, march on with their forces; so that I wish your men would march all night in good order, lest they be surprised; here, they shall have all the accommodation the place will afford; in this hurry, pardon me for this brevity, the rest, the bearer will inform you.

I rest, Sir, Your faithful Servant,

Robert Lundy.

If the men be not landed, let them land, and march immediately.

* Derriana, p. 32.

1.

1825.

2.

1831.

3.

4

1855.

MAJOR-GENERAL BRICE, C.B.

(From Ensign, 1848, to Colonel of the Regiment, 1903.)

(2)

Sir,

Since the writing of this, Major Tiffin is come here, and I have given him my opinion fully, which, I believe, when you hear and see the place, you will both join with me, that without an immediate supply of money and provisions, this place must fall very soon into the enemy's hands. If you do not send your men here some time to morrow, it will not be in your power to bring them at all. Till we discourse the matter,

I remain, dear Sir,

Your most faithful Servant,

R. L.

Accordingly, Governor Lundy ordered Colonels Cunningham and Richards, to leave their men on board their ships, and to come with some of their officers to the town, that they might resolve on what was to be done.

THE MILITARY COUNCIL.

April 16th. Colonels Cunningham and Richards, with some of their officers, came to the town, where Colonel Lundy called a council of war, composed of 16 persons.

Colonel Lundy. Lord Blaney. Colonel Cunningham. Colonel Richards.	Lieut.-Colonel T. Hussey. Captain R. Echlin. „ J. Tranter. „ J. Lyndon.	Officers of Colonel Cunningham's Regiment.
Captain Cornwall (commander of the frigate " Swallow.") Major Tiffin, " The Queen's " Regiment. Captain James Hamilton. Mr. Moggridge, Town Clerk.	„ J. Taylor. „ C. Coote. Capt.-Lieut. D. Pearson. *Lieutenant H. Pache (Paget) ?	Officers of Colonel Richards' Regiment.

The two colonels, with their officers, were entire strangers to the town, and the rest were, in a great measure, unacquainted with it, except Mr. Moggridge; and when several of the principal officers of the town, who had some suspicion of Colonel Lundy's design (such as Colonels Francis Hamilton, Chichester, Crofton, Ponsonby, &c.), desired to be admitted, they were absolutely

* (It is doubtful whether this was an officer of Colonel Richards' regiment or a civilian.)

refused, though, at the same council, he (Colonel Lundy) pretended he had sent for the two first named, but said they could not be found, and Sir Arthur Rawdon, he had said, was dying. Colonel Cunningham delivered to Governor Lundy His Majesty's letter and orders, but the Governor, as president of the council, gave the same account of the state of the town as he had before given to Major Tiffin, and therefore advised them all to quit it, as he said he would do himself. Those of the council who came from England, thinking it impossible that the Governor should be ignorant of the condition of the town, and observing the account to pass without any contradiction, by those who had been there some time, but had not, it appears, informed themselves better, soon agreed in the opinion of returning to England, rather than stay in a place, not to be victualled, especially when he, (the Governor), had said the enemy were near their gates with 25,000 men, and there was no possibility of a return from England in so short a time.

ONLY COLONEL RICHARDS URGED AGAINST IT,

" because he looked on deserting that garrison, not only as " quitting the city, but the whole kingdom."*

Accordingly the Military Council came to the following

RESOLUTION.

Upon inquiry, it appears,—That there is not Provision in the Garrison of Londonderry, for the present Garrison, and the two Regiments on Board, for above a Week, or Ten Days at most, and, it appearing, that the Place is not tenable against a well-appointed Army ;—Therefore it is concluded upon, and resolved ; That, it is not convenient for His Majesties Service, but the contrary, to land the two regiments under Colonel Cunningham and Colonel Richards, their Command now on Board in the River of Lough-foyle.

That, considering the present Circumstances of Affairs, and the likelihood, the Enemy will soon possess themselves of this place, it is thought most convenient, that the principal Officers shall privately withdraw themselves, as well for their own preservation, as in hopes, that the Inhabitants, by a timely Capitulation, may make terms the better with the

* Walker's " Siege of Derry."

Enemy; and, that, this we Judge most convenient for His Majesties Service as the present State of Affairs now is.———

As the result of this Council, Colonels Cunningham and Richards, with their officers, went down to the ships, which, that day, were below Redcastle. But Colonel Lundy, to delude both officers and soldiers in the town, (who were earnestly begging that the English forces might land), told them publicly: "It was resolved, the English forces should immediately "land, and when they were in their quarters, the gates should "be opened, and all join in the defence of the town."*

And to cloak the intrigue better, the sheriffs were ordered to provide quarters for them, which they accordingly did. But this was all mere sham to amuse the town, whilst they, (the newly arrived troops), might get away with greater ease and safety.

When the result of the proceedings of the Council of War became known, it occasioned great uneasiness and disorder in the town, which had very ill effects upon the Governor and his Council, who, finding that they could not be further serviceable, thought fit to retire.

Governor Lundy could not so easily make his escape, being conceived more obnoxious than any of the rest, and found it convenient to keep to his chamber.

A council being appointed, the Revd. George Walker and Major Baker meeting him there, desired him to continue his government, and that he might be assured of all the assistance they could give him, but he positively refused to concern himself any further.

Finding him desirous to escape the danger of such a tumult, they suffered him to disguise himself, and, in a sally, for the relief of Culmore, to pass in a Boat, with a load of Match on his back, from whence, he got to the Shipping.

The Garrison, seeing they were deserted, and left without a Governor, and having resolved to maintain the town and defend it against the enemy, unanimously resolved to choose Mr. Walker and Major Baker to be their Governors during the Siege; but these Gentlemen considering the importance, as well as the uncertainty of such an office, acquainted, by letter, Colonel Cunningham (whose business they thought it

* Derriana, p. 37.

was to take care of them), with this matter, and desired him to undertake the charge.

His answer was: "That he, being ordered to apply "himself to Colonel Lundy, for direction in all things, relating "to their Majesties Service, could receive no application from "any that opposed that authority."

On the 18th April the ships moved to Green Castle, and on the 19th sailed for England.

On the 1st May, an order was issued that Colonels Cunningham and Richards were immediately to attend the King at Hampton Court, which they did, when His Majesty was so displeased at the representations of the Governor of Londonderry not having been sufficiently investigated, and that the Governor's suggestions had been yielded to, when there was reason to doubt his integrity, that he deprived the two colonels of their commissions.

The Calendar of State Papers, Domestic Series, dated 1689–90, page 18, shows the following:—

"*May 24th*, 1689. Warrant to William Sharpe, Messenger, to deliver Colonel Cunningham and Colonel Richards, prisoners, into the custody of the keeper of the Gatehouse"; so that these two unfortunate officers were imprisoned in the Tower, along with Colonel Lundy (who had been brought from Ireland), and presumably, they remained in custody until August 12th, the date on which the proceedings of the House of Commons Committee, (which assembled on 3rd June), were brought before the House.

Copy of Minutes, from the Journal of the House of Commons.

Proceedings of the Committee which Assembled to Enquire into the State of Londonderry, and why Colonels Cunningham and Richards had Returned.

1689. June 3rd. Resolved.

That His Majesty be humbly desired to give directions, that copies of the Commissions and Instructions given, relating to Londonderry, and the Kingdom of Ireland, be transmitted to the Committee, and, also, that Colonel Lundy, a prisoner in the Tower, may be brought before it.

The following evidence was taken* :—

SIR ARTHUR ROYDEN stated: That, three days before the two regiments arrived from England, Colonel Lundy had told him, that there was then in the town, three months' provisions for 6000 men.

CORNET NICHOLSON said: That, there was at this time, a great store of provisions in the town, every house having plenty; that provisions came daily in boats, sufficient for 3000 men, as Colonel Lundy himself had told Lord Blaney a little while before the Council of War was held.

LORD BLANEY said: That, there was a proposition made, to destroy all the ammunition left in the town, which the Governor approved, saying it was better than letting it fall into the enemy's hands, but nothing was resolved in this matter.

COLONEL CHICHESTER said: That, that afternoon, Colonels Cunningham and Richards, and most of the gentry and officers present at the Council of War, went down to the ships, as the people thought, to bring up the men, but, when the people saw the ships had drifted lower down the town, they became alarmed, and cried out, that they were betrayed; also, that the Governor (Colonel Lundy) said the Council of War had resolved, that the men should be landed, and, to make it more creditable, pretended to give some orders about quarters; and, when so many gentry, going down to the ships, alarmed the townspeople, he said, they only went to see the men land. Colonel Chichester further informed the Committee, that Captain Cornwall, Commander of the "Swallow," frigate, (which took Colonel Cunningham to Londonderry), had, on his return voyage, brought a great many Protestant passengers on board his ship, and demanded £4 a head from each, failing the receipt of which, he plundered them of their swords, watches, clothes, or anything they had, in a very barbarous manner.

COLONEL LUNDY, being several times examined, says: That, Major Tiffin, when he brought Colonel Cunningham's letter, told him, that they, (the two regiments,) had brought no provisions for the town, and proposed that Colonel

* John Mackenzie, p. 400.

Cunningham might come and discourse with him before the men were landed, to which he consented.

COLONEL CUNNINGHAM says he gave Tiffin no such order.

COLONEL LUNDY owns the proceedings at the Council of War, and says, he did not know but what provisions were scarce, as he had represented them. He denies the several discourses and confessions, which the witnesses have charged him with.

COLONEL CUNNINGHAM, on being examined, admits the proceedings of the Council of War, but denies having said: "he would go home, let who would be displeased"; he denies, that his brother ever came to the ships, only Captain Cole, and, that he himself, having a good opinion of Colonel Lundy, bid Captain Cole go back, and obey the Governor.

HOUSE OF COMMONS, dated, August 12th, 1689.

Sir Thomas Littleton reports, from the committee, appointed to enquire into the miscarriages, relating to Ireland and Londonderry:*

THAT, the committee had examined several witnesses, and came to no resolution thereon, but had directed him, to report the matter specially to the House, as to how they found the same, viz.—

It appeared to the committee:

THAT, the same day, the fight was at Cladyford, (April 15th,) Colonels Cunningham and Richards arrived in Londonderry River, with the two regiments;

THAT, Colonel Cunningham wrote two letters to Colonel Lundy, the import of which was to acquaint him of his arrival, and to know, in what condition the town stood, and receiving no answer, he sent Major Tiffin to the Governor, with a third letter. Then, Major Tiffin meeting Colonel Lundy's messenger, carrying an answer to the two former letters, he, (Major Tiffin,) took him back to Londonderry, where Colonel Lundy opened his letter and inserted a postscript, with the particulars already mentioned.

*John Hempton, "Siege of Derry," p. 381.

THAT, the next morning, Colonel Cunningham sent to Colonel Richards, to bring three or four of his officers with him, and he, taking the like number of his own, they all went to Londonderry, leaving the men on board the ships.

THAT, they went directly to the Governor's house, where they met a great number of the gentry and officers that were then in the town, and Colonel Richards said, that Colonel Cunningham and the Governor whispered together at the window, but he knew not what they said.

THAT, in a short time after arrival, Colonel Lundy proposed to go to a Council of War in the Council Chamber, whither they went accordingly.

THAT, Colonel Lundy had refused to admit divers of the officers, who used to attend former councils, particularly Colonels Hamilton and Chichester, Majors (Revd.) Walker and Baker, and, that Colonel Chichester and Major Walker, endeavouring to enter, Colonel Lundy ordered them to be kept out, saying it was to be a select company. Colonel Lundy was the first to speak at the Council, and made a proposition, to quit the town, and send the two regiments back again, alleging as his reason, that there was not above a week, or ten days' provision left in the town, and that the enemy, (King James's army,) was 25,000 strong, and, within four or five miles.

THAT, this proposition was consented to by all present, without any contradiction, except that COLONEL RICHARDS OPPOSED IT, saying that "Quitting the "town was quitting of a kingdom," whereupon one present, arose and swore he would be hanged for no man's pleasure, which he, (Colonel Lundy) thinks was Major Tiffin, and another said, "he would go home, let "who would be displeased," which he thinks might be Colonel Cunningham, but is not positive.

THAT, all present, signed a paper of consent, to quit the town, which paper began in these words: "Finding, "upon enquiry, that there was not above a week, or "ten days' provision," &c., whereas in truth, no enquiry was made at the Council of War, but all present took the Governor's word for it, who informed them, that he had searched the stores.

THAT, they all agreed afterwards, upon their honour, not to discover what resolution they had taken; Colonel Richards says, the Governor proposed an oath of secrecy, which he was the first to refuse, and it was rejected.

THAT, when the ships had drifted so far away, it appeared plain, there was no intention of landing men; the people then went and beset Colonel Lundy's house, and, from that time, watched him so closely, that he could not make his escape to the ships, which stayed for him—yet, that he sent to Colonel Cunningham not to go away without him, lest he became a sacrifice to the rabble.

THAT, the next morning, (April 17th,) Mr. Cunningham, the Colonel's brother, and Captain Cole were sent from the town to the ships, to Colonel Cunningham, to offer him the government of the town, if he would come and take possession of it, and land his two regiments, at the same time stating they had sufficient provisions in the town, and that they being in great numbers, would march out, and leave the garrison to his care. But Colonel Cunningham refused the offer, bidding them go back and obey their governor, all of which he confessed, except that Captain Cole was the only messenger sent to him.

THAT, the ships returned to Liverpool, with most of the officers and gentry belonging to the town, but Colonel Lundy was left behind at Londonderry, whence, he afterwards escaped to Scotland in a private soldier's clothes.

UPON COLONEL RICHARDS BEING EXAMINED, it appeared to the Committee, THAT COLONEL CUNNINGHAM WAS HIS COMMANDER-IN-CHIEF, AND THAT HE HAD ONLY ACTED THROUGHOUT, IN OBEDIENCE TO HIS SUPERIOR OFFICER.

Sir Thomas Littleton acquainted the House, that he was directed by the said Committee, to move the House, that an address may be presented to His Majesty, that the said Colonel Cunningham may be allowed bail.

The Committee of the House of Commons RESOLVED :—

THAT, an humble address be presented to His Majesty, by the Members of the Privy Council, that Colonel Lundy be sent to Londonderry to be tried, for the treasons that are laid to his charge.

THE HOUSE ADJOURNED.

It may be here noted, that King William III. (prior to the return of the two regiments) had, apparently, placed such faith in Colonel Lundy's loyalty, that he had appointed him, (as Governor of Londonderry,) to raise a regiment of foot in the service of Ireland, (dated April 20th, 1689,) Colonel Lundy himself being its colonel.

The Calendar of State Papers, Domestic Series, " William " and Mary," dated, October 7th, 1689, states :—

OPINION, by Lord Massarene, Lord Kingston, Colonels Philips and Crofton, Dr. Walker, Mr. Davis, and Mr. Rowley :—

" THAT, at present, it is neither safe, nor practicable, to send " Colonel Lundy to Ireland, to be tried."

* * * * * * * * * *

From the foregoing short narrative, it will be seen that Colonel Solomon Richards was most anxious to avert the catastrophe of the two regiments returning, but was overruled, and if the decision of the Committee of the House of Commons, (absolving him from blame), had only been awaited, it seems highly probable that King William would, at least, have adopted a less stern measure, than to have deprived him of his commission—a most deplorable occurrence.

Colonel Richards was, however, honoured at his death two years later, (6th October, 1691), by being buried in Westminster Abbey, (North Cloister).

BATTLE OF SHERIFFMUIR.

The following is a true copy of an account of the Battle of Sheriffmuir, written by Major-General Wightman, Colonel of the regiment at the time.*

Sterling. November 14, 1715, at 11 at night.

Last Friday, I arrived from Edinburgh, where, I had finished all the Works and Barricadoes, that I had Orders to do for the Security of that Town; and, as soon as I came to his Grace the Duke of Argyle, he told me he was glad to see me, and that as he intended to make a march towards the Enemy, the next morning, he had sent an Express to Edinburgh for me. Accordingly on Saturday, the 12th Instant, our whole Army marched over the Bridge of Sterling towards the Enemy, who lay at a Place called Ardoch, about seven Miles from this Place, and, in the Evening, our Army came within about three Miles of the Enemy's Camp.

We lay all that night on our Arms, and the next Morning being Sunday, by break of Day, I went with his Grace, where our advanced Guard was posted, and had a plain View of the Rebels Army, all drawn up in Line of Battle, which consisted of Nine thousand and one hundred Men. They seemed to make a Motion towards us: Upon which, the Duke ordered me immediately back, to put our Men in Order; and soon after his Grace ordered them to March to the top of a Hill against the Enemy: But before all, or not above half our Army was form'd in Line of Battle, the Enemy attack'd us. The Right of their Line, which vastly out wing'd us, lay in a hollow Way, which was not perceiv'd by us, nor possible for us to know it, the Enemy having Possession of the Brow of the Hill; but the Left of their Army was very plain to our View. The moment we got to the top of the Hill, not above half of our Men were come up, or could form; the Enemy, that were within little more than Pistol-shot, began the Attack with all their Left upon our Right. I had the Command of the Foot;

* Patten's History of the Rebellion.

The Enemy were Highlanders ; and as it is their Custom, gave us Fire ; and a great many came up to our Noses, Sword in hand ; but the Horse on our Right, with the constant Fire of the Plottoons [platoons] of Foot, soon put the Left of their Army to the Rout. The Duke of Argyle pursuing, as he thought, the Main of their Army, which he drove before him above a Mile and a half over a River. As I march'd after him as fast as I could, with a little above three Regiments of Foot, I heard great Firing on our Left ; and sent my Aid-de-Camp to see the Occasion of it, and found that the Right of the Enemy's Army, that lay in the hollow Way, and was superior to that Part of their Army which we had beaten, was fallen upon the Left of our Line, with all the Envy imaginable ; and, as our Men were not form'd, they cut off just the half of our Foot, and our Squadrons on our Left. The Duke, who pursued the Enemy very fast, was not appriz'd of this ; And as he had order'd me to march as fast as I could after him, I was obliged to slacken my March, and send to his Grace to inform him of what had happened.

I kept what Foot I had in perfect Order, not knowing but my Rear might soon be attack'd by the Enemy that had beat our Left, which proved to be the Flower of their Army. At last, when the Duke had put to Flight that part of the Rebel-Army he was engaged with, he came back to me, and could not have imagin'd to see such an Army as was behind us, being three times our Number ; but as I had kept that part of our Foot which first engag'd in very good Order, his Grace join'd me with five Squadrons of Dragoons, and we put the best Face on the Matter to the Right about, and so march'd to the Enemy, who had defeated all the Left of our Army. If they had had either Courage or Conduct, they might have entirely destroy'd my Body of Foot ; but it pleased God to the contrary.

I am apt to conjecture, their Spirits were not a little dampt, by having been Witnesses some Hours before of the firm Behaviour of my Foot, and thought it hardly possible to break us.

We march'd in a Line of Battle, till we came within half a Mile of the Enemy, and found them ranged at the top of a Hill, on very Advantageous Ground, where their Horse could not well attack us ; For we had the Convenience of some

Earth-Walls, or Ditches, about Breast high; and, as Evening grew on, we inclined with our Right towards the Town of Dunblain, in all the Order that was possible. The Enemy behaved like civil Gentlemen, and let us do what we pleased; so that we passed the Bridge of Dunblain, posted ourselves very securely, and lay on our Arms all Night. This Morning we went with a Body of Dragoons to the Field of Battle, brought off the wounded Men, and came to this Town in the Evening.

General Webb's late Regiment, now Morrison's (8th), is one of the unfortunate Regiments that were not formed, and suffered most. . . .

General Evans had his Horse shot under him, and escaped very narrowly, as well as myself.

P.S. Our whole Army did not consist of above a Thousand Dragoons, and Two Thousand Five Hundred Foot; and but a little more than Half of them engaged. However, I must do the Enemy that Justice to say, *I never saw Regular Troops more exactly drawn up in Line of Battle, and that in a Moment; and their Officers behaved with all the Gallantry imaginable. All I can say, is, it will be of the last Danger to the Government, if we have not Force to destroy them soon.* The Loss on both Sides I leave for another Time, when we have a more exact Account.

MAJOR-GENERAL UTTERSON, C.B.

(From Ensign, 1854, to Colonel of the Regiment, 1905.)

Centre-piece, Officers' Mess, 1st Battalion.

REGIMENTAL PLATE, PICTURES, &c.

Although the 17th Regiment was raised in 1688, the officers' mess plate of the 1st Battalion only dates back to about April, 1840, owing to the disaster which befel the regiment, in having lost its mess plate in March, 1840, in the wreck of the transport " Hannah " off the coast of Bombay.

It is much to be regretted that there has been no means of even obtaining particulars of any of the lost plate, so as to have enabled facsimiles of it, in some instances, to have been reproduced.

After this severe loss, a certain amount of necessary plate had to be at once purchased regimentally, the Government of India having refused to sanction any compensation to the mess for losses incurred by the shipwreck. (See Appendix 6.)

In the early fifties of the nineteenth century, the practice of presenting some article on promotion, or on leaving the regiment, came into vogue, with the result that the officers' mess of the 1st Battalion since 1853, and that of the 2nd Battalion since 1859 (it having been raised in 1858), have become the recipients of many very handsome cups and other useful articles.

In 1890, a centre-piece (height nearly 4 feet 2 inches), was purchased for the mess of the 1st Battalion (to celebrate the bi-centenary of the regiment two years previously), which, for excellence in workmanship and design, would be hard to surpass, and must be seen to be appreciated at its true value. It was manufactured by the Goldsmiths' and Silversmiths' Company, of Regent Street, by whose courtesy the accompanying illustration is produced.

The following is a list of presentation plate of the 1st Battalion :—

Claret jug, 1853. Presented by Lieutenant W. P. Williams.
Sevastopol salver, 1854. Presented by Captain Lindesay.
Cup, 1863. Presented by Lieutenant H. Burnett.
Bread basket, 1864 and 1889. Presented by Officers, 2nd Battalion "Queen's."
Cup, 1865. Presented by Lieutenant J. H. Thorold.
Cup, 1867. Presented by Major W. Earle.
Cup, 1868. Presented by Subalterns, 1st Battalion 17th.
Cup, 1868. Presented by Lieut. S. Hobson. " With Polly Hobson's love."

Cup, 1869. Presented by Colonel W. Gordon.
Ram's head snuff box, 1877. Presented by Surgeon Ruxton.
Cup, 1877. Presented by Mrs. F. G. Lacon, "In Memoriam" Lt. F. G. Lacon
Tea set (6 pieces and salver), 1879. Presented by Colonel G. T. Brice.
Cup, 1887. Presented by Major Nares, "In Memory of."
Silver inkstand, 1887. Presented by Colonel A. H. Utterson.
Centre-piece, 1890. Purchased from mess fund.
Bell, 1894. Presented by Colonel W. M. Rolph.
Bowl, 1896. Presented by Lt.-Col. Alderson and Officers, Mounted Infantry.
Claret jug and 2 Cups, 1897. Presented by Lieut.-Colonel Vulliamy.
Cup, 1901. Presented by Captain W. H. Young.
Cigarette box, 1902. Presented by Captain R. Selby, R.A.M.C.
Cup, 1905. Presented by Lieutenant S. Hamilton, R.E.
Elephant's foot (liqueur stand), 1905. Presented by Lieutenant H. M. Travers.
Cup, 1906. Presented by Lieutenant H. A. Hildebrand.
Cup, 1907. Presented by Lieutenant C. S. Davies.
Bridge box, 1908. Presented by Major Tarry.
Medal menu holders (3), 1908. Presented by Colonel Burne.
Napkin rings (18), 1908. Presented by Captain G. Rolph.
Silver match box, 1909. Presented by Captain J. Moore, R.E.
Do. 1909. Presented by Captain H. Mackworth, D.S.O., R.E.
Do. 1909. Presented by Lieutenant G. Sim, R.E.
Do. 1909. Presented by Lieutenant D. Hogg, R.E.
Silver snuff boxes (4). No inscription on any.
One do., with four compartments.
Barometer, 1910. Presented by Captain F. I. Ford.
Silver cigarette box, 1911. Presented by Captain J. Bacchus.
Cup, 1911. Presented by Captain C. S. Davies.

The following cups are also in the possession of the battalion :—

1890.—Band Cup. Presented by the Halifax-Bermuda Cable Company.
1890.—Governor's Cup, Bermuda Steeple-chase. Won by "Blizzard."
1900.—Cape Hunt Cup.
1903.—Madras Football Cup.
1907.—Lydd Rifle Meeting Cup. Won by "A" Company.
1907–08.—Tennis Cup. Won by Major Blackader and Captain Challenor.
1908.—Miniature Army Rugby Cup.
1910.—Machine Gun Cup. Won by machine gun team.
1911.—Miniature Army Rugby Cup.
1911.—Machine Gun Cup. Won by machine gun team.
1912.—Miniature Army Rugby Cup.

Other articles of interest are the massive candelabra, purchased out of the mess fund some years ago ; a very old snuff box, with four compartments in it (mentioned above), and the inscription illegible from age, believed to have been the only relic saved from the wreck of the "Hannah," in March, 1840. Another of these snuff boxes (about 8 by 4 inches and 1½ inches deep) has on the lid a raised figure of a tiger recumbent, surrounded by a wreath and "XVII" engraved.

It was given to the mess by officers of the detachment at Aden (out of the mess profits), shortly after the wreck of the "Hannah."

Also a handsome drum-major's staff, surmounted by a silver tiger, presented by Major G. I. Walsh in 1908.

In the same year, the following articles were given by Lt.-Col. Webb (retired), viz. :—

A small frame, containing old patterns of lace worn in the ranks from 1768 to 1836, and of drummers' lace and fringe, as worn by the drummers of the regiment, from about 1820 to 1871.

A case of nine shoulder-belt plates as worn by the officers of the 17th from the earliest period to the date of their abolition in April, 1855, including four actually worn by officers of the regiment.*

1. Small silver oval, 1776.
2. Silver oval, larger size, 1st Battalion, 1799—1815.
3. Silver oval, 2nd Battalion, 1799—1802 (when the battalion was disbanded.)
4. Dead gilt, oblong, 1815—25 (12th Oct.).
5. Silver oblong, 1825 (13th October)—1828.
6. Same pattern, in bright burnished gilt, 1829-30.
7. Dead gilt, oblong, smaller size, 1831—28th September, 1843.
8. Bright gilt, oblong, 29th September 1843 to 2nd October, 1845.
9. Gilt matted, oblong, 3rd October, 1845 to 31st March 1855.

In September, 1908, a case of 17 medals, all won by 17th men, was given by Colonel G. H. P. Burne, on retirement, five of which were in pairs, showing the obverse and reverse of them. They comprised :

The Army of India Medal (single), with clasps for Nepaul and Bhurtpoor.
Medal for Ghuznee (single).
Crimean Medal (single), with clasp for Sevastopol
The Turkish (Crimean) Medal (single).
Fenian Raid, Canada, 1866 (a pair).
Humane Society (single).
Good Conduct Medal (single).
Afghan Medal, 1878–80 (a pair), with clasp for Ali Masjid.
Burma Medal, 1887–89 (a pair).
Queen's South African (a pair), with clasps for Talana, Defence of Ladysmith, Laing's Nek and Belfast.
King's South African Medal (a pair), with clasps for South Africa, 1901 and 1902.
Medal (single), given by the Officers, "For Military Merit" in 1811.

When the battalion was at Aldershot, in 1910, a Queen's South African Chocolate Box, of the year 1900, was "By

* No. 4 worn by Capt. George Peevor, and presented by his grandson, Leslie Peevor, Esqre.
Nos. 7 and 9 worn by Capt. Edw. Croker, and presented by his son, Lt.-Col. H. L. Croker.
No. 8 worn by Capt. Alex. M'Kinstry, and presented by his daughters.

"command of His Majesty King Edward VII., presented "by the Army Council to the 1st Battalion Leicestershire "Regiment, as one of the battalions which served longest "in the South African War, 1899–1902."

A book of photographs with descriptions and ribbons of various medals. Presented by Lieut.-Colonel Brind, 1912.

Amongst the pictures are some interesting old prints and a water-colour painting, showing types of old uniforms worn by the regiment in the years 1742, 1800, 1830 (four after E. Hull), and the Band 1861–70, given by Lt.-Col. Webb. Others have been given as follows :—

SUBJECT—

"Death of General Wolfe." Presented by Major Martin-Martin, R.E., 1889.
Queen Victoria. Purchased.
Queen Alexandra. Purchased.
King Edward VII. Purchased.
Major-General the Honourable R. Monckton. Purchased.
Portrait of Colonel Alexander M'Kinstry. Presented by Colonel M'Kinstry.
General Tytler. Presented by General Tytler, V.C., C.B.
Recruiting Notice. Presented by Captain G. Pearson, 15th Hussars.
General Wolfe. Presented by Colonel G. H. P. Burne, 1908.
Wellington and Nelson. Presented by Colonel G. H. P. Burne, 1908.
Foot Soldier (George III.). Presented by Colonel G. H. P. Burne, 1908.
Two Topographical prints of Waterloo. Presented by Col. Burne, 1908.
"Death of General Montcalm." Presented by Lieut.-Colonel Sherer, 1909.
Twelve Lithographs of the siege and capture of Ghuznee and Khelat. Purchased.
Bombardment of Sevastopol. Purchased.
Panorama of the siege of Sevastopol. Purchased.
Coloured print: The Siege of Huy. Purchased.
Charge of the Dragoons at Gravelotte. Purchased.
Portrait of His Majesty King George V. Presented by Lt.-Col. H. L. Croker, 1911.
Officers in the Crimea, 1855. Presented by Lt.-Col. Webb, 1911.
Two prints, Sevastopol. Presented by Lieut. V. M. Cooper, R.N., 1912.

2ND BATTALION.

The presentation plate of the Officers' Mess, dates from its formation on the 24th March, 1858, and is as follows :—

Tigress's head (snuff box). Presented on the formation of the corps, by Lieut.-Colonel Crofton and 29 officers.
Claret jug, 1859. Presented by Ensigns Braddell, Bros, H. Burnett, Caird, Dwyer, Deane and Elgin, to commemorate the presentation of Colours by Lady Vivian.
Claret jug, 1859. Presented by Lieutenant H. S. Wedderburn.
Cup, 1859. Presented by Lieutenant H. Burnett.

Snuff box, 1859. Presented by Lieutenant Versturme.
Cup, 1864. Presented by Officers, 2nd "Queen's."
Carriage clock, 1868. Presented by Captain H. S. Wedderburn.
Candelabra (3), 1871. Purchased.
Cup, 1872. Purchased, in remembrance of the Beagles, 1870–72.
Cup, 1871. Presented by Lieutenant Bird.
Silver toothpick case, 1872. Presented by Lieutenant S. L. Richards.
Cup, 1872. Presented by Lieutenant F. Blackley.
Silver fire holder, 1872. Presented by Officers, 61st Regiment.
Cup, 1873. Presented by Lieutenant E. Allfrey.
Silver inkstand, 1874. Presented by Lieutenant E. Allfrey.
Indian coffee salver, 1879. Presented by Major-General Brice.
Centre-piece, 1880. Presented by Lieut.-Colonel J. B. H. Boyd.
Silver (Indian) claret jugs (2), 1881. Presented by Major Dunning and Captain F. F. Parkinson.
Salver, 1881. Presented by Captain F. F. Parkinson.
Silver (Cashmir) stand, for passing round cigars and cigarettes, 1881. Presented by Captain W. Gregg.
Milk jug and sugar bowl, 1884. Presented by Captain E. R. Scott.
Twenty goblets, 1880—1889. Presented by Earl Howe, and won as follows:—

Two cups,	1880.	Won by	Captain H. I. Nares.
,,	1881.	,,	Lieutenant A. W. M'Kinstry.
,,	1882.	,,	Lieutenant G. D. Carleton.
,,	1883.	,,	Captain R. J. G. Creed.
,,	1884.	,,	Lieutenant G. D. Carleton.
,,	1885.	,,	Major C. Middlemass.
,,	1886.	,,	Captain R. J. G. Creed.
,,	1887.	,,	Captain V. Bunbury.
,,	1888.	,,	Captain J. Mosse.
,,	1889.	,,	Lieutenant H. Gordon.

Sugar bowl and milk jug, 1886. Presented by Captain Flood, 2nd "Queen's"; Captain Forbes, 6th Royal Regiment; Lieutenant Thomas, 31st Regiment.
Elephant's foot (liqueur stand), 1890. Presented by Major G. D. Carleton. (Elephant shot in Somaliland.)
Tiger's head (snuff box), 1890. Presented by the Maharajah Cooch Behar.
Silver hand bell, 1893. Presented by the Subalterns, 2nd Battalion.
up, 1894. Presented by Colonel C. F. W. Moir.
Cup, 1894. Presented by 1st Battalion Scots Guards.
Silver candlesticks (pair), 1894. Presented by Lieutenant Mignon.
Silver barometer, 1896. Presented by Captain F. H. Alexander.
Silver cigar box, 1897. Presented by Captain C. Hunt.
Silver cigarette box, 1897. Presented by Officers of "A" and "B" Squadrons, 17th Lancers.
Silver hand bell, 1898. Presented by Major B. G. Humfrey.
Silver bridge box, 1901. Presented by Lieutenant Ford.
Silver bowl, 1902. Presented by Major Lawrie, R.E.
Silver match box, 1903. Presented by Lieutenant E. Henderson.
Silver match box. Presented by 2nd Lieut. Durnford, 2nd West India Regt.
Silver cigar cutter, 1903. Presented by Captain F. Gruchy.
Silver inkstand, 1905. Purchased.

Silver bridge box, 1905. Presented by Officers 1st Battalion King's Own Scottish Borderers.

Cup, 1908. Presented by Lieutenant Bacchus.

Coffee pot, 1908. Presented by Captain C. G. Liddell.

Indian silver (snake) fire holder. Purchased.

The following cups are also in the possession of the battalion :—

1887–88. Silver Bowl, Bengal Presidency Rifle Association, Inter-regimental. Won by regimental team.

1887–88. Cup, Bengal Presidency Rifle Association, Inter-regimental. Won by regimental team.

1905. Cup, Eastern District Bronze Medal Tournament, Officers' Bayonet Team combats. Won by regimental team of five officers.

1906. Cup, Colchester Preliminary Tournament.

1906. Cup, Colchester Preliminary Tournament, Officers' Bayonet Team combat. Won by regimental team of five officers.

1906–11. Cup. Record of the challenge cups won by the battalion, when stationed at Belgaum, from October to March in these years. (Subscribed for.)

1910. Cup. Poona Divisional Assault-at-arms. Best regiment at arms.

The following cups and shields have been purchased regimentally :—

Inter-company Rugby football cup; Inter-company hockey cup; Inter-company Association football shield; Inter-company cricket shield; Inter-company shooting shield.

Also a case of 12 medals, mostly won by 17th men, which comprise :—

Army of India Medal, 1799—1826, and clasp for Nepaul. } Given by
Medal for Louisburg, 1758. } Lt.-Col. Brind.
Ghuznee Medal, 1839.
British Crimean Medal (2), and Clasp for Sevastopol.
Sardinian Crimean Medal.
French Crimean Medal.
Fenian Raid, Canada, 1866.
Afghan Medal, 1878–80 (2), and Clasp for Ali Masjid.
Burma Medal, 1887–89.
Medal for Distinguished Conduct in the Field.

Other articles of interest are: A miniature of General Wolfe, of which there is apparently no record* ; a small frame containing old patterns of the drummers' lace and fringe, as worn by the drummers of the regiment, from about 1820 to 1871, given by Lt.-Col. Webb in 1906.

Amongst the pictures are many of interest, including one of Queen Victoria and the Prince Consort, presented by Her Majesty the Queen in commemoration of a review of troops,

* A copy on ivory of a well-known miniature of this distinguished General. Presented to the Mess by the late Captain Logan, and said to have been painted by Captain Logan's sister.

1st January, 1891, at which the battalion was present, it being then quartered at Warley; also an oil colour, of Aden, painted and presented by Major G. D. Carleton. Others have been given as follows:—

SUBJECT—

His Majesty King Edward VII. Presented by Major H. L. Croker.
Her Majesty Queen Victoria. Purchased.
H.R.H. the Prince of Wales, 1875. Purchased.
H.R.H. the Duke of Cambridge. Purchased.
The Emperor Napoleon I. Purchased.
The Duke of Wellington. Purchased.
Garden Party, Tower of London. Purchased.
Colonel Alexander M'Kinstry. Presented by Captain A. W. M'Kinstry.
Major F. W. Reader. Presented by Major Reader.
Captain R. H. Dunning. Presented by Captain Dunning.
Captain W. S. Copland. Presented by Captain Copland.
Lieutenant H. J. F. Balmain. "In Memoriam." Presented by his sister.
Types of the 17th Foot. Presented by Colonel C. F. W. Moir.
Do. (1830). Presented by Lt.-Col. Webb.
H.M.S. "Duncan," and Officers (2). Presented by Officers, H.M.S. "Duncan," 1866.
Louisburg. Presented by Lt.-Col. Brind.
"Death of General Wolfe." No record (very old).
Wellington and Nelson. No record (very old).
Battle of the Boyne. No record.

SERGEANTS' MESS, 1ST BATTALION.

The sergeants of the battalion possess a good deal of handsome plate, some of which has been won by them at shooting and at various sports, and some has been presented. Their collection dates from 1872, since which time (and especially of recent years) every effort has been made to win all competitions open to them. Amongst their prizes are some interesting shields and other presentations, and some of the sergeants have also presented pleasant souvenirs. In their mess can also be seen some very interesting pictures and portraits, which are well cared for.

The following is a list of the plate:—

Silver Cup (Shooting), 1872, 96th Regiment. Won by Sergeants' Mess Team.
Silver Cup, 1878. Presented by Mr. T. Whimper, Indian Brewery.
Silver-mounted Clock, 1879. Presented by Brigdr.-Genl. A. H. Cobbe, C.B.
Inkstand, 1886. Presented by Sergeants, Royal Marine Artillery.
Marine Clock, 1886. Presented by Sergeants, Royal Marine Light Infantry.
Silver Cup (Shooting), 1890, Royal Engineers. Won by Sergeants' Mess Team.
Silver Cup (Shooting), 1890, Army and Navy Rifle Meeting, Bermuda. Won by Sergeants' Mess Team.
Shield, Regimental (Shooting), 1891. Presented by Lt.-Col. W. M. Rolph.

Shield, Regimental (Football and Cricket), 1891. Presented by Lieut.-Colone W. M. Rolph.

Silver Challenge Cup (Shooting), 1898. Presented by Lieut.-Colonel F. W. Reader.

Silver Horse Shoe, (Felix), 1901. Presented by Lieut.-Colonel G. D. Carleton.

Shield, Regimental (Hockey), 1901. Presented by Major G. H. P. Burne.

Silver Cup (Boxing), 1903, All India Light Weights. Presented by Sergeant F. Cunningham.

Silver Cup (Boxing), 1908, Light Weights, Canterbury. Presented by Sergeant J. Sanderson.

Silver Cup (Football), miniature replica, Army Rugby Challenge. Presented to the winning team.

Case for cigars and cigarettes (Shooting), 1910. Won by Sergeants' Mess Team.

Silver Cup, 1910. Presented by Sergeants, 1st Battalion Royal Irish Fusiliers (87th).

In 1909, an aneroid barometer was won at the Hythe Rifle Meeting by the Sergeants' Mess Rifle Team, and in the same year the mess was presented with a writing desk by the sergeants of the 1st Battalion Royal West Kent Regiment (old 50th).

For the football year 1909–10, the battalion also held the following silver cups, won by the regimental football teams :—

(*a*) Aldershot Senior Football Challenge Cup.
(*b*) Six-a-side Football Challenge Cup.

SERGEANTS' MESS, 2ND BATTALION.

The collection of the sergeants' mess plate of the 2nd Battalion dates from 1887, and they also possess some very handsome plate, mostly shooting prizes, including challenge cups and other presents from officers, which are combined with trophies won in running and football competitions, and by the sergeants' bayonet teams.

When quartered at Belgaum (1906–11) there was a large scope for their zeal, in the Divisional rifle meetings, and Divisional Assault-at-arms, held at Poona, where they acquired some beautiful trophies, won after hard-fought victories.

They also possess some very interesting pictures and portraits.

The following is a list of the plate :—

Silver Cup, 1887, 1st Prize, Sergeants' Sprint Race.

Silver Cup, Regimental Challenge (Shooting), 1895. Presented by Lieut.-Colonel Gregg.

Silver Cup, Recruits' Sectional (Shooting), 1896. Presented by Colour-Sergt. Beckett.

Silver Inkstand, 1896. Presented by Sergeants, 1st Battalion Royal Scots.

Silver Tankard, 1897 (Shooting). Presented by Sergt.-Instructor Gibson.
Silver Tankard, 1899 (Shooting). Presented by Sergt.-Major Hammond.
Silver Tankard, 1899 (Shooting). Presented by Sergeant Daft.
Silver Tray, 1899. Presented by Sergeants, 1st Battalion Argyll and Sutherland Highlanders.
Silver Cup, Recruits' Sectional (Shooting), 1899. Won by Col.-Sergeant Pursey.
Silver Cup, Queen Victoria's (Shooting), 1902. Won by Regimental Team.
Silver Cup, Cairo (Shooting), 1902. Won by Regimental Team.
Silver Cup, Queen Victoria's (Shooting), 1903. Won by Regimental Team.
Silver Cup, Recruits' Sectional (Shooting), 1903. Won by Sergeant Toon.
Silver Cup, Queen Victoria's (Shooting), 1904. Won by Regimental Team.
Silver Cups (2), miniature (Football), 1904. Presented by Captain H. Logan.
Silver Cup, Recruits' Sectional (Shooting), 1905. Presented by Lieut.-Colonel Scott.
Silver Cup, Eastern Division Athletic Meeting, 1906. Won by Sergeants' Bayonet Team.
Silver Cup, 6th Poona Division Assault-at-arms, 1908. Won by Sergeants' Bayonet Team.
Silver Cup, Southern Mahratta Railway Rifles (Shooting), 1908. Won by Regimental Team.
Silver Cup, Championship Trophy, best British unit (Shooting), 1908–09. Won by Regimental Team.
Silver Cup, Defence (Shooting), 1909. Won by Regimental Team.
Silver Cup, Southern India Rifle Association (Shooting), 1909. Won by Regimental Team.
Shield, Sergeants' Running, 1909. Presented by members of the Mess.
Silver Cup, Non-commissioned Officers (Shooting), 1909–10. Won by Non-commissioned Officers, 2nd Battalion.
Silver Cup, Non-commissioned Officers' Plate (Shooting), 1910. Presented by the Officers. Won by "H" Company.

In 1909, a shield of the Royal Army Temperance Association was presented to the mess by its members, and in the same year a cross-country running cup by Colonel V. T. Bunbury, D.S.O.

CLAIMS FOR COMPENSATION FOR LOSSES SUSTAINED BY THE 17TH REGIMENT, IN THE WRECK OF THE TRANSPORT "HANNAH," IN MARCH, 1840.

PRICES OF REGIMENTAL CLOTHING LOST OR DAMAGED.

	Rs.	A.	P.
3 Staff Sergeants' Coats at Rs. 16	48	0	0
3 Pairs Staff Sergeants' Trowsers, at Rs. 9	27	0	0
2 Do. Sergeants' do. at Rs. 5	10	0	0
321 Privates' Coats, at Rs. 5.3.2	1668	8	6
50 Pairs Ammunition Boots, at Rs. 3.3.2	159	14	4
Total Rs.	1913	6	10

LIST OF REGIMENTAL NECESSARIES LOST FROM THE QUARTERMASTER'S STORES.

	Price of each.			Total amount.		
	R.	A.	P.	R.	A.	P.
100 brass brushes	0	4	9	29	11	0
100 shoe do.	0	10	4	64	9	3
107 button do.	0	3	9	25	1	3
100 cloth do.	0	6	3	39	1	0
32 forage caps	0	15	8	31	5	4
138 stocks and clasps	0	9	6	81	15	0
120 knives and forks	0	6	0	45	0	0
177 spoons	0	1	6	16	9	6
70 razors	0	7	3	31	11	6
108 pairs of braces	0	9	0	60	12	0
107 towels	0	6	2	41	3	10
86 number pieces, grenadier caps	0	3	2	17	0	4
73 do. light company	0	3	2	14	7	2
100 do. battalion	0	3	8	22	14	8
126 yards Privates' red cloth	3	1	8	391	2	0
60 pairs country boots	1	5	0	78	12	0
Total Rs.				991	3	10

The following was claimed for the loss of two band instruments :—

One valve trumpet—Rs.	100
One cornet ,,	80
Total Rs.	180

The compensation claimed by the Officers' Mess for the loss of mess baggage, and camp equipage, was as follows :—

Mess plate, plated ware, knives and forks, &c.	Rs. 1318
The whole of the camp equipage	,, 550

Copy of letter from the Secretary to the Government of Bombay to the Military Auditor-General, dated, Bombay Castle, 13th May, 1840 :—

"I am directed to acknowledge the receipt of your letter of the 25th ultimo (No. 124), and, in reply, to acquaint you that the Honourable the Governor in Council sanctions your passing the bills prepared by the officers and men of Her Majesty's 17th Regiment for compensation for loss of baggage by the wreck of the 'Hannah.'"

The following was the decision of the Governor-General in Council, dated, Fort William, 27th May, 1840, in reply to the Government of Bombay :—

"I am directed to state for the information of the Government of Bombay, that the Right Honourable the Governor-General of India in Council is pleased to sanction the grant of compensation for loss of necessaries sustained by H.M.'s 17th Regiment, owing to the wreck of the ship 'Hannah,' to the amount of Rs. 991.3.10, according to Statement No. 4 of the documents accompanying your despatch.

With regard to the claims made on account of clothing, the property of the colonel, mess property and band instruments, His Lordship in Council is unable to authorise the grants solicited, compensation for losses of those descriptions being wholly unknown to the regulations of the Service.

(Signed)

Secretary to the Govt. of India, Military Department.'

It should be mentioned that as a slight set off to the losses by this shipwreck, all ranks of the regiment had already received batta money for the previous year's campaign in Afghanistan, and though it was submitted in the correspondence, that this emolument was hardly intended to cover unforeseen losses by shipwreck, the ruling as above proved unfavourable.

THE REGIMENTAL CHAPEL, ST. MARTIN'S CHURCH, LEICESTER.

THIS Memorial Chapel was instituted in 1897 by Colonel W. M. Rolph, commanding the 17th Regimental District, Leicester, after consultation with officers commanding 1st and 2nd Battalions of the regiment, in conjunction with the Rev. Canon Sanders, M.A., LL.D., by whose courtesy the following particulars concerning it have been supplied.

Visitors to the Church of St. Martin are always interested as they enter by the south porch, to find themselves in front of the tattered colours of the 17th Regiment, and the stained-glass windows and brasses which have been erected in memory of its officers and men. And, many are they who come specially to see these objects of interest, and to read the numerous names inscribed on the monuments which have accumulated so quickly. One can scarcely be in the church a few minutes without finding someone who wants to visit this part of it, specially, to find in some cases, names of relatives or friends, whose memorials they desire to see, or to look once more at the flags under which they themselves served.

This portion of the church dates from about A.D. 1260. It is younger than the rest of the building, the greater portion of which was built about thirty years earlier.

The Regimental Chapel was formerly dedicated to St. George, the Patron Saint of England, and was built at the same time as the great South Aisle, the eastern portion of which (though now a Consistory Court) was dedicated to our Lady. The priests' stalls still remain. The whole aisle was under the management and control of the Guild of Corpus Christi, who had so much to do with shaping the government of the town in the early days. The chapel was divided off by a parclose screen from the rest of the aisle. Entering through this screen, one would behold an altar, over which was a "vowle," or canopy, and at the back of it hung a painted cloth. The window was coloured, and the jambs painted in blue and red, fragments of which colours still remain. In the chapel was a series of stalls, probably for the use of the Master and Stewards of the Guild, curiously carved. North says: "One of them had a projecting bracket or "'miserere' on the under side, which, when turned up for use, "exhibited in bold carving a 'dragon or flying serpent, "'with long talons and expanded wings' of a black leaden "colour, under which were two human skulls." Through the sitting board of this stall was a hole large enough to admit a thick wand or badge of office during the ceremonies of the Guild.

But by far the most conspicuous object in the chapel, if not in the whole church, was the great figure of St. George on horseback, which was, as Throsby says, "harnessed in the "Church splendour of the times," and which stood on a platform of some height. This, placed on a trolley, formed a conspicuous object in the yearly procession of the Guild through the streets of the town.

All this splendour disappeared at the time of the Reformation, and is now only a matter of history. During the intervening years we believe the chapel served as an organ loft, to hold a gallery for school children, and finally as a vestry. And it is only in the last few years that it has been dedicated to its present use. This came about in the following way. Some officers from Glen Parva came to the Vicar to express their desire that the colours which were placed in various parts of the church, with the monuments relating to the Crimean, Afghan, and Burmese campaigns, and to Bermuda, might be collected together, and placed in some corner which they might look upon as a Regimental Corner, and where future monuments might be accumulated. The Vicar (Canon Sanders) gladly fell in with the idea, and suggested St. George's Chapel. The idea of again associating the warrior's spirit and the old idea of conflict and domination of good over evil with the modern soldier, appealed to his mind; and with pleasure he offered them this "corner" for their purposes. The design was carried into effect, and first of all the old colours and monuments were placed there, the old colours mounted on netting, so that they would no longer drop to pieces, and then the monuments began to be erected.

On this date (31st March, 1912) there are four sets of colours deposited, viz. :—

1st Battalion	Those issued in	1830;
	,,	1854;
2nd ,,	,,	1859;
1st ,,	,,	1879 (but not taken into use until about 1882).

NOTE.—The Colours were originally placed in the nave against the pillar immediately behind the present pulpit; they were subsequently moved and placed on either side of the great west door, being eventually placed in their present position on the formation of the present Memorial Corner.

The following memorials are also in the Regimental Chapel :—

(i) The Crimean Memorial, erected by the Officers, Non-Commissioned Officers, and Privates of the 17th Regiment, to the memory of comrades who died in the service of their country during the Crimean Campaign, 1854 to 1856. Capt. John L. Croker, Lieut. L. J. Seagram, Surgeon W. Simpson, and 210 Non-Commissioned Officers and Men. This was originally placed at the west end of the church in 1859.

(ii) Large brass tablet in memory of Officers and Soldiers of the 1st Battalion 17th Foot who lost their lives in Afghanistan during the campaign 1878–9. Capt. J. H. Gamble, Lieuts. N. C. Wiseman, C. G. Whitby, and E. Allfrey; 2nd Lieut. E. H. Watson, and 43 Non-Commissioned Officers and Men.

(iii) Brass tablet to the memory of Officers, Non-Commissioned Officers and Men of the 2nd Battalion who were killed in action or died in Burma during 1888 and 1889. Capt. G. F. Shaw, Lieut. J. H. F. Balmain, and 25 Non-Commissioned Officers and Men.

(iv) Brass tablet to the memory of Non-Commissioned Officers and Men 1st Battalion, who died in Bermuda and at Halifax, N.S., September 1888—April 1891 (34 names).

(v) Brass tablet in memory of Lieut. Arthur Cleghorn Thomson, who was killed in action at Bida, W. Africa, on January 26th, 1897.

(vi) Brass tablet in memory of Major E. H. Peacock, who died at St. Helena on the 16th of July, 1897.

(vii) To the memory of Officers, Non-Commissioned Officers and Men, 2nd Battalion, who died in Egypt during the years 1900—1902. Captains W. S. Copland, and C. M. P. Hassall, and 40 Non-Commissioned Officers and Men.

(viii) In loving memory of Lieut.-Colonel F. F. Parkinson, A.P.D., and formerly Captain 17th Regiment, who died at Singapore, 30th May, 1902. Erected by his widow and son.

(ix) To the memory of Lieut.-Colonel J. G. L. Burnett, commanding 1st Battalion Leicestershire Regiment, who died at sea on the 9th September, 1904, and is buried at Aden.

(x) To the memory of Major-General Geo. Tito Brice, C.B., Colonel of the Leicestershire Regiment, who died on September 21st, 1905.

(xi) Tablet to the memory of Capt. H. S. Logan, who died 9th May, 1908, from wounds received in action, whilst commanding a detachment of Soudanese troops near Mesellamia, Soudan, on May 1st, 1908.

There are two special windows :—

(1) In memory of the late Lieut.-Colonel F. W. Reader, who died in Natal, November 10th, 1898. This window has three panels containing figures representative of St. John, St. George, and St. Martin.

(2) The window commemorating the services of four battalions of the Leicestershire Regiment (including Militia and Volunteers), who were killed in action or died of wounds, or disease, in the South African War, 1899—1902. Its marble tablets below contain the names of Captain H. C. Thorold, Lieutenants W. M. J. Hannah and C. P. Russell, and 128 Non-Commissioned Officers and Men.

It may be described as of five lights, with numerous tracery openings, of the 14th century or " Decorated " character.

The lights are of equal dimensions, and are now filled with stained glass of military interest.

In the centre light is a figure of Our Lord in Resurrection as the Victor Mortis, bearing the banner of Triumph.

In a small panel beneath, the two angels of the Sepulchre are represented bearing tablets inscribed with the words : " He is not here for He is risen as He said."

In each of the other four lights is represented a Biblical military character, these being Joshua and David as warriors of the Old Testament, followed by the Centurion as referring to the early Christian Church, and St. George, symbolic of later times, and the Saint in whose name the Regimental Chapel was dedicated.

Beneath the figure of Joshua is the scene of his command to the Sun to "stand still." Under that of David is the incident of his victory over Goliath. In connection with the figure of Cornelius is the incident of his Baptism, while St. George's triumph over the Dragon completes the series.

In the opening at the Vertex of the window is displayed the Badge of the Regiment. Immediately contiguous are introduced angels bearing olive branches and palms.

At a lower level, within four openings are the words—Angel borne—"Thanks be to God which giveth us the victory." Then follow the names of the engagements of the Campaign in which the Regiment fought with such distinction, Talana, Defence of Ladysmith, Laing's Nek, Belfast, Cape Colony, Orange Free State, Transvaal, Natal, South Africa, 1899–1902. Within the canopies above the principal figures, Angels bear scrolls, bearing respectively the words "Watch Ye," "Stand fast in the Faith," "Be Strong."

Since 1905, it has been ruled by the officers of the 1st and 2nd Battalions, that the only memorials, in future, to be erected in the Regimental Chapel, are to be to officers who have been killed or died on field service, or from wounds received in action.

Of course, improvements are still wanted, the floor should be of marble, and once more, the chapel should be enclosed by a screen, the walls of which should afford space which will be wanted on which to affix more memorials. Meanwhile, the Regiment has a "corner" which it may be proud of, for its position in the Central Church, its historical associations, and its present condition.

ST. GEORGE'S CHAPEL,
ST. MARTIN'S CHURCH, LEICESTER.

Containing the Colours and Memorials of the Leicestershire Regiment.

By the courtesy of Messrs. Raphael Tuck & Sons, Ltd.

THREE PRESENTATION CASES TO THE REGIMENT, OF ITS ORNAMENTS, BUTTONS, LACE, &c., FROM 1765 TO 1910. (FROM LT.-COL. WEBB.)

CONTENTS OF THE THREE PRESENTATION CASES.

(1765–1910.)

LEFT-HAND CASE.

Shako Plates (12) of officers and rank and file from 1800 (the earliest date), to 1871; three of universal pattern, the remainder regimental; also patterns of chains worn.

Helmet Plates (8), with patterns of chains, of officers and rank and file, from 1879 to 1910.

Forage Cap Badges (19) of officers and rank and file from 1830 to 1910.

Puggaree Badges (4) of officers and rank and file from 1870 to 1889.

Skirt Ornaments (2) worn by the rank and file of the Grenadier and Light Companies, of the Crimean period. (Those of the battalion companies wore a button only.

Officers' Collar Ornaments (2), worn in full and service dress respectively. Introduced since 1871.

Band Pouch Ornament, adopted since 1881.

Bridle Ornaments (2) of universal pattern, for Infantry mounted officers, from 1831 to 1902.

Buttons, of regimental and territorial patterns, of officers and rank and file, from 1767 to 1910.

Badges of rank, gold and silver, embroidered and metal, dispersed at intervals.

CENTRE CASE.

Officer's Grenadier Cap Plate, of universal pattern, worn from 1765 to 1799.

Do. do. 1831 to 1845.

Pair of Undress Scales, worn by Captain Alexander M'Kinstry, Grenadier Company, with the single-breasted blue frock coat, 1834–45.

Pair of Full Dress Epaulets, as worn by battalion company officers, 1831–55.

Officer's Gorget, of universal pattern, as worn in the reign of King George IV.

Double Whistle and Chain, worn by officers of the Light Company, 1831–55.

Pouch Ornament, worn for two years only by officers of the Light Company, 1855–57.

Coatee Skirt Ornaments, Battalion Company officer, 1820 and 1838.

Do. Grenadier and Light Company officers, 1838.

Colour-Sergeant's Badge, worn before 1868.

Do. In present use (1910).

Pair of Full Dress Wings (one each), as worn by officers of the Grenadier and Light Companies respectively, 1831–55.

Silver Breastplates, three oval and one oblong, 1776–1815.

Gilt do., five oblong, 1816 to 31st March, 1855.

Waistbelt Plates, 11 patterns, as worn by officers and rank and file, 1825–1910.

RIGHT-HAND CASE.

Lace, Patterns of gold and silver lace, worn by officers and sergeants from 1822 to 1910. (By warrant officers, from 1st July, 1881.)

Pattern of worsted lace, worn by rank and file to 1836, and special pattern by drummers to 1871.

Officers' Shoulder Cords, from 1831 to 1910.

Shoulder straps, worn by rank and file, 1857–1910, and metal titles, &c.

Old Net Sash, worn round the waist by sergeants, 1824–45.

The following badges, worn on the right arm, introduced since 1871, some of which are worn in 1910, viz.:—

- Pioneers,
- Judging Distance,
- Armourer Sergeants,
- Quartermaster Sergeants,
- Signallers,
- Drummers,
- Bandmaster and bandsmen.

Scouts' Badges, 1907–10.

Badges for skill at arms:

- (*a*) Swordsmanship.
- (*b*) Musketry.

Each case measures 4 feet by 1 foot 7 inches.

REGIMENTAL MUSIC.

"Wolfe's Dirge," consists merely of a few bars as a lament to the great hero, General James Wolfe, and is played by the band on parade as the "Officers' Call," by both battalions of the regiment.

The slow marches, which were adopted for a long time by both battalions, appear to have been lost sight of, since marching past in slow time has for some years been practically abolished. That of the 1st Battalion in 1853 is said to have been "Robin Adair," and later, "The Bird of the Desert." That of the 2nd Battalion was known by the name of "The Grenadiers."

There is in possession of the 1st Battalion an old "march past" air, in quick time, known merely as "1772," which is typical of the style of old English airs of that period.

From the date of its formation in 1858, the 2nd Battalion for many years used to march past to "The Warwickshire Lads," but both battalions now go by to the "Romaika," a Grecian air, said to have been introduced into the 1st Battalion by a medical officer who was posted to it from the 64th Regiment. The "Romaika" was published by authority as the official march past in 1882.

On guest nights the officers' "Mess Calls" of both battalions are played on cornets, the first of these being "A stag will die this day," and the second "The Roast Beef of Old England."

As the finale to the band programmes on guest nights, the bands of both battalions play the old Leicestershire hunting air, "We'll all go a-hunting to-day," "1772," and "Romaika."

"WOLFE'S DIRGE."

ROMAIKA.

INDEX.

K.

L

M

* Exclusive of R.A., R.E., Cavalry, Royal Marines, &c.

* Exclusive of R.A., R.E., Cavalry, Royal Marines, &c.

Y

www.ingramcontent.com/pod-product-compliance
Ingram Content Group UK Ltd.
Pitfield, Milton Keynes, MK11 3LW, UK
UKHW021840270726
14058UKWH00002B/250